Markt- und Unternehmensentwicklung / Markets and Organisations

Edited by
A. Picot, München, Germany
R. Reichwald, Leipzig, Germany
E. Franck, Zürich, Switzerland
K. M. Möslein, Erlangen-Nürnberg, Germany

Change of institutions, technology and competition drives the interplay of markets and organisations. The scientific series 'Markets and Organisations' addresses a magnitude of related questions, presents theoretic and empirical findings and discusses related concepts and models.

Edited by

Professor Dr. Dres. h. c. Arnold Picot
Ludwig-Maximilians-Universität
München, Germany

Professor Dr. Professor h. c. Dr. h. c.
Ralf Reichwald
HHL – Leipzig Graduate School
of Management, Leipzig, Germany

Professor Dr. Egon Franck
Universität Zürich, Switzerland

Professorin Dr. Kathrin M. Möslein
Universität Erlangen-Nürnberg,
Germany,
HHL – Leipzig Graduate School
of Management, Leipzig, Germany

Bastian Bansemir

Organizational Innovation Communities

 Springer Gabler

RESEARCH

Bastian Bansemir
Universität Erlangen-Nürnberg
Nürnberg, Germany

Dissertation Friedrich-Alexander-Universität Erlangen-Nürnberg, 2011

ISBN 978-3-658-01301-1   ISBN 978-3-658-01302-8 (eBook)
DOI 10.1007/978-3-658-01302-8

The Deutsche Nationalbibliothek lists this publication in the Deutsche Nationalbibliografie; detailed bibliographic data are available in the Internet at http://dnb.d-nb.de.

Library of Congress Control Number: 2013932215

Springer Gabler
© Springer Fachmedien Wiesbaden 2013

Printed on acid-free paper

Springer Gabler is a brand of Springer DE.
Springer DE is part of Springer Science+Business Media.
www.springer-gabler.de

For us

Foreword

20 years ago, Linus Torvalds revolutionized the way services and products are innovated: What began with publishing software code (i.e., the operating system "Linux"), the vivid development of a community of programming experts and hobbyists alike, still fascinates and inspires practitioners and researchers. Many phenomena which we nowadays discuss under terms such as "Open Source", "Open Innovation", and "Wisdom of the Crowds" may be rooted back to the development of Linux over the past 20 years. Since then, community-based innovation endeavors have fundamentally changed the way we work, collaborate, innovate, and even the way we spend our spare time.

While we continuously engage in, and take advantage of community-based innovations mostly without even recognizing, community-based interaction and innovation within organizations is still a challenge. Organizations often struggle with the question of how to efficiently and effectively deploy community-based innovation mechanics within their organizational boundaries to increase (1) speed, (2) quantity, and (3) quality of innovation development. Bastian Bansemir addresses this challenge and focuses his research on organizational innovation communities. He convincingly reports in the thesis at hand surprising, solid, and fascinating insights from his scholarly journey.

Bastian Bansemir's work invites the reader to explore:

- a systematic and comprehensive literature review of organizational communities and the implications for organizational innovation communities,

- twelve case studies from varying industry sectors that provide an in-depth understanding concerning the influence of organizational contexts,

- key triggers to initiate, support, foster, and execute community-based innovation development in organizations based on longitudinal, and in-depth action research, and

- the role of "self-efficacy" (self-confidence) and "positive affect" (positive emotions) to drive knowledge exchange as a precondition for innovation to occur in organizational innovation communities.

The thesis offers comprehensive knowledge, solid, interesting, and surprising insights, and lays a sound foundation for further research to be conducted. It provides comprehensive overviews of current research streams for students enthusiastic to gain knowledge of the topic. The insights provided include contra-intuitive findings which deserve further research. Practitioners will profit from extremely helpful recommendations on how to set up organizational innovation communities.

Research by Bastian Bansemir has already profited from presentations and feedback at national and international conferences, e.g. the Research Colloquium on Innovation and Value Creation, the International Conference on Business Informatics, the International Product Development Conference (IPDMC), the European Academy of Management Conference (EURAM) and the Annual Conference of the Academy of Management (AoM). His work has been published in journals such as International Journal of Knowledge-based Organizations and BuR - Business Research.

The work at hand appeals by its theoretical reach and empirical scope, the fresh approach and attractive presentation. It has been accepted as doctoral dissertation in 2011 by the School of Business and Economics at the University of Erlangen-Nuremberg. The book deserves broad dissemination both in the research community and in management practice. It is especially recommended to those with a deep interest in driving interactive innovation in organizations.

I wish you, Bastian, all the best for your future career!

Prof. Dr. Kathrin M. Möslein

Acknowledgements

The thesis at hand originates from my activities as a research associate at the Chair of Information Systems I – Innovation & Value Creation (Prof. Dr. Kathrin M. Möslein) at the Friedrich-Alexander-University Erlangen-Nürnberg (2007 to 2011) within the project "Open-I: Open Innovation within the Firm."[1] I am very grateful, and happy about the ongoing support that I experienced while authoring this book.

First, I would like to thank *Kathrin Möslein* for her encouragement, support, feedback, trust, and unbridled optimism. I gladly remember discussions, and collaboration with you, especially while building-up the chair in 2007 and 2008. I would also like to express my thanks to *Michael Amberg* for authoring the second opinion.

Second, I would like to thank the chair team *"nah und fern"* for supporting me when I needed it, challenging me whenever it was possible, cultivating bad habits, just having good times or sharing bad times. Specifically, I would like to thank *Doctor (Anne-Katrin) Neyer* for your extremely helpful and always challenging professional advice, and your moral support along the way. This work would not have reached this quality without you. Thank you, *Jörg Haller, Vivek Velamuri, Babsi (Barbara Inmann)*, and *Marga Stein, Christiane Rau, Sabrina Adamczyk, Dominik Böhler, Jens Söldner*, and *Stefan Hallerstede* for many unforgettable hours. You made my stay in Nürnberg.

Third, I would like to express my thanks to the members of the project "Open-I", to all professors, post-docs and doctoral candidate (e.g., *Michael Reinhardt*) for giving me a great time while working on the project. Specifically, I would like to share my appreciation with Datev and Flughafen München for their engagement. For me *Steffen Henne, Brigitte Fischer, Gabriele Ilg, Stefan Häberlein*, and *Claudia Donig* played an important role to identify exciting insights, to achieve very good project results, and to make this experience a good one.

Fourth, I was able to build this thesis on competencies, and knowledge gained during my former education at the Technical University Munich, and the Michaeli-Gymnasium-München. I would like to thank all peers, teachers (e.g., *Karl-Heinz Kern*), lecturers, and professors (e.g., *Ralf Reichwald*) who supported me in the one or the other way.

Fifth, special thanks go to me friends who had to stand my absence from Munich, my struggles, but also my euphoria in the last couple years. Thank you, dear *Andreas Lademann, Simona Kindlein*, the crew of Brett-à-Porter (*Matthias Steiner, Oliver Zimmer, Vivien Dollinger, Matthias Grimmeisen, …*) and "The girls" (*Katharina Vogl, Franziska Schmahl* etc. and of course the corresponding boys, such as *Bernd Haas* and *Boris Großkopf*).

Sixth, my family also deserves a big thank you for supporting me all the years in multiple ways and on various occasions: My parents *Gudrun* and *Horst*, as well as my brothers *Axel* and *Gero*, but also my "parents-in-law" *Anna* and *Robert-Peer Hintelmann*.

Finally, my beloved and longtime significant other *Anna-Katharina Hintelmann* deserves a very big THANK YOU for everything along the way. You know best for what, but you may want to choose from encouragement, pressure when necessary, sympathy when needed, support, help, and tolerance when asked for, compensation when I was stressed or consulting when I was confused, etc.

---

[1] Funded by the Federal Ministry of Education and Research (BMBF; FKZ: 01FM07054)

Overview of contents

Part I   Introduction 1

   1   Relevance...2

   2   Structure 6

   3   Research paradigm 9

Part II   Systematic literature review...13

   1   Setting the stage..14

   2   Organizational communities: A literature review..18

   3   Implications for organizational innovation communities.......................................50

Part III A taxonomy of organizational innovation communities

         (empirical study I)...53

   1   Setting the stage..54

   2   Theoretical perspectives 56

   3   Research methods...58

   4   Findings 62

   5   Discussion 77

   6   Conclusion 80

Part IV Sensemaking in organizational innovation communities

         (empirical study II) 83

   1   Setting the stage..84

   2   Theoretical perspectives 85

   3   Research methods...90

   4   Findings 96

   5   Discussion 107

   6   Conclusion 111

Part V   Knowledge exchange in organizational innovation communities

         (empirical study III)...113

   1   Setting the stage..114

   2   Theoretical perspectives 116

   3   Research methods...120

4   Findings 124

5   Discussion 130

6   Conclusion 134

Part VI   Reflection 135

1   Summary and contribution 136

2   Practical implications 142

3   Research implications 145

Table of contents

Part I   Introduction 1

  1   Relevance 2

  2   Structure 6

  3   Research paradigm 9

Part II   Systematic literature review 13

  1   Setting the stage 14
    1.1   A brief history and constraints of community literature 14
    1.2   Defining organizational innovation communities 15
    1.3   Towards a community framework: Input, mediator and outcome 17

  2   Organizational communities: A literature review 18
    2.1   Research methods 18
    2.2   Input factors 19
    2.3   Mediators 31
    2.4   Outcomes 41
    2.5   Discussion: Organizational communities 42
    2.6   Discussion: Boundary-less communities 46

  3   Implications for organizational innovation communities 50

**Part III   A taxonomy of organizational innovation communities
            (empirical study I)** 53

  1   Setting the stage 54

  2   Theoretical perspectives 56
    2.1   Organizational integration 57
    2.2   Transition strategies 57

  3   Research methods 58
    3.1   Sample and data collection 58
    3.2   Data analysis 60

  4   Findings 62
    4.1   Influence of organizational integration 62
    4.2   Transition strategies 71

  5   Discussion 77

  6   Conclusion 80

**Part IV   Sensemaking in organizational innovation communities
            (empirical study II)** 83

1   Setting the stage 84

2   Theoretical perspectives 85
2.1   Major means of innovation development 86
2.2   Sensemaking as an interpretive framework 87

3   Research methods 90
3.1   Research design 90
3.2   Sample and data collection 91
3.3   Data analysis 94

4   Findings 96
4.1   Coherence 96
4.2   Cognition 99
4.3   Collaboration 101
4.4   Outcomes 104

5   Discussion 107
5.1   Nested coherence 107
5.2   Collective flow 108
5.3   Collaborative transformation 109

6   Conclusion 111

Part V   Knowledge exchange in organizational innovation communities
          (empirical study III) 113

1   Setting the stage 114

2   Theoretical perspectives 116
2.1   Self-efficacy and knowledge exchange 117
2.2   Affective states and knowledge exchange 118

3   Research methods 120
3.1   Research design 120
3.2   Manipulations and data collection 121

4   Findings 124
4.1   Manipulation checks 124
4.2   Test of hypotheses 125

5   Discussion 130

6   Conclusion 134

Part VI   Reflection 135

1   Summary and contribution 136

2   Practical implications 142

3   Research implications 145

List of figures

Figure 1:   Structure of the thesis 6
Figure 2:   The starting page of the Open-I platform 10
Figure 3:   Input, mediator, outcome framework (based on Mathieu et al., 2008a, p. 413) ... 17
Figure 4:   Example of using Atlas.ti for data analysis 61
Figure 5:   Four forms of organizational integration, based on transition strategies 78
Figure 6:   The social process of sensemaking 89
Figure 7:   Screenshot of the virtual whiteboard used for virtual workshops and
            observations 93
Figure 8:   The social process of sensemaking for innovation 107
Figure 9:   Influence of self-efficacy on knowledge exchange between control and
            experimental groups 126
Figure 10: Influence of self-efficacy on knowledge exchange between control and
            experimental groups based on platform-based indicator of knowledge exchange
            126
Figure 11: The influence of positive affect on knowledge exchange between control and
            experimental groups 128
Figure 12: The influence of self-efficacy on knowledge exchange between control and
            experimental groups based on platform-based indicator of knowledge exchange
            128
Figure 13: Innovation development on the Open-I platform 172
Figure 14: Innovation evaluation on the Open-I platform 173

List of tables

Table 1: Tom 3
Table 2: Ina 4
Table 3: Norm 5
Table 4: Organizational context: Strategy, structure, culture, and ICT 20
Table 5: Publications assigned to organizational context 23
Table 6: Community context: Boundary objects and seeding patterns......................... 24
Table 7: Publications assigned to community context.. 27
Table 8: Member factors: Motivation, trust and status 28
Table 9: Publications assigned to member inputs... 31
Table 10: Emergent states: Proximity, strength of ties, centrality and stickiness.......... 33
Table 11: Publications assigned to emergent states 35
Table 12: Processes: knowledge, learning and social processes..................................... 37
Table 13: Publications assigned to processes 40
Table 14: Outcomes: Qualitative and quantitative approaches....................................... 41
Table 15: Publications assigned to outcome... 42
Table 16: Summary of organizational community literature 46
Table 17: Summary of boundary-crossing community literature 49
Table 18: Crucial gap in research concerning organizational integration 51
Table 19: Crucial gap in research concerning social processes...................................... 51
Table 20: Crucial gap in research concerning engagement 52
Table 21: Taxonomy: Existing literature, crucial gap and insights............................... 54
Table 22: Theoretical background: Organizational integration and transition strategies 56
Table 23: Case study companies... 59
Table 24: Dyadic integration: Definition, findings, learning, and project example 63
Table 25: Quotations concerning dyadic integration 64
Table 26: Cultural integration: Definition, findings, learning and project example............... 65
Table 27: Quotations concerning cultural integration 67
Table 28: Structural integration: Definition, findings, learning, and project example 68
Table 29: Quotations concerning structural integration.. 69
Table 30: No integration: Definition, findings, and learning.. 70
Table 31: Quotations concerning neither cultural nor structural integration.......................... 71
Table 32: Initiating transition strategy: Definition, findings, learning, and project example.. 72
Table 33: Quotations concerning initiation strategy 72
Table 34: Negotiating transition strategy: Definition, findings, learning, and
          project example.. 73
Table 35: Quotations concerning negotiation strategy 74
Table 36: Narration transition strategy: Definition, findings, learning, and project example . 75
Table 37: Quotations concerning narration strategy... 76
Table 38: Taxonomy: Summary of part III 80
Table 39: Sensemaking: Existing literature, crucial gap, and insights 84
Table 40: Theoretical background: Constraints of innovation development and
          sensemaking.. 86
Table 41: Overview of data sources.. 94
Table 42: Summary of aggregated dimensions, second-order constructs, and first-order
          concepts 96
Table 43: Coherence: Definition, findings, learning, and project example 97
Table 44: Quotations concerning coherence... 98
Table 45: Cognition: Definition, findings, learning, and project example 99

Table 46: Quotations concerning cognition..101
Table 47: Collaboration: Definition, findings, learning, and project example...................102
Table 48: Quotations concerning collaboration...104
Table 49: Outcomes: Definition, findings, and learning...104
Table 50: Quotations concerning outcomes...106
Table 51: Sensemaking: Summary of part IV..111
Table 52: Knowledge exchange: Existing literature, crucial gap, and learning 114
Table 53: Theoretical background: Self-efficacy and positive affect and
          knowledge exchange..116
Table 54: Self-efficacy and positive effect: Positive influence on knowledge exchange......125
Table 55: Summary of results of study 1 127
Table 56: Summary of results of study 2 129
Table 57: Knowledge exchange: Summary of part V..134
Table 58: Summary of part II...137
Table 59: Summary of part III 138
Table 60: Summary of part IV 139
Table 61: Summary of part V 140
Table 62: Toolbox to configure organizational innovation communities 144
Table 63: Publications assigned to organizational context 174
Table 64: Publications assigned to community context...175
Table 65: Publications assigned to individual inputs..176
Table 66: Publications assigned to processes 177
Table 67: Publications assigned to emergent states..179
Table 68: Publications assigned to outcomes 180

Part I Introduction

1 Relevance

This year Linux celebrates its 20th anniversary.[2] What makes this event interesting for this dissertation is that Linux, along with other initiatives like Apache or Mozilla, fundamentally changed our understanding of how innovations are created. Until 20 years ago, innovation development mostly took place in designated research and development (R&D) departments (Rothwell, 1994), which did not differ significantly from what Thomas Edison established at Menlo Park in 1876. Linux was one of the first organizations to deviate from this traditional approach of developing innovations in designated R&D departments by relying on innovation communities instead (Lerner & Tirole, 2002). In contrast to the hierarchical order and control structures present in R&D departments, they are characterized by democratic structures, in the sense that *community members engage mutually (voluntarily on the basis of reciprocity) in innovation development while sharing similar repertoire and objectives.* Even experienced contributors, like Eric Raymond, were skeptical about innovation development within such communities (Raymond, 1999, pp. 2–3). However, they have often outperformed traditional innovation development (Amin & Roberts, 2008; Barab, MaKinster, & Scheckler, 2003; Bogenrieder & Nooteboom, 2004; Lindkvist, 2005; Plaskoff, 2003; Wenger, Mc Dermott, & Snyder, 2002; Wenger, 2000; Brown & Duguid, 1991; Raymond, 1999; Lakhani, Jeppesen, Lohse, & Panetta, 2007). For instance, Lakhani et al. (2007) are able to show that innovation communities are superior in terms of quantity, speed and quality. Hence, it is not surprising that organizations increasingly rely on innovation communities in the pursuit of innovation (Brown & Duguid, 1991, p. 44; Sawhney & Prandelli, 2000; Neyer, Bullinger, & Möslein, 2009; Richter, Mörl, & Koch, 2011). The recent trend of applying innovation communities as a new paradigm of innovation development can be observed in manifold occurrences. However, even after 20 years of innovations created by Linux, organizations still struggle to use innovation communities as an approach to develop innovations within their boundaries, i.e. to successfully anchor, organize and foster organizational innovation communities.

To describe these challenges a somewhat unusual approach is chosen: The perspectives of Tom, a top-level manager, Ina, an innovation manger, and Norm, a 'normal' employee are subsequently displayed to narrow down key challenges and to show the relevance of the thesis.

Tom is around 45 years old, studied business administration at a renowned German university, and started his career as a business consultant. After more than a decade of consultancy, he joined a large company as a board member.[3] In his spare time he plays golf to network with former colleagues, potential business partners and thought leaders.

This is what he experiences: He is fascinated by the many examples of organizational innovation communities and their application (i) in a wide spectrum of industries and (ii) in organizations of varying sizes. For the former, examples of Cisco's I-Zone and Swarovski's i-flash communities come to mind: Whereas Cisco's I-Zone community frequently develops innovative services; Swarovski's i-flash community generates product-related innovation. For the latter, Tom thinks of Datev, with around 5,000 employees, currently introducing the DIP-community (Datev Innovation Pool) but also of IBM's innovation jam (with IBM having more than 400,000 employees on the payroll). Moreover, he is also convinced about the potential of organizational innovation communities to not only develop (i) huge amounts of innovation but also (ii) in a variety of areas. Tom has Daimler's Business Innovation Community in mind: a respectable number of around 1,500 business ideas have been created

---

[2] For a brief history of Linux, a short film is available at http://video.linux.com/video/2225.
[3] See profile below.

from 2008 to date. Moreover, his friends at Siemens report about TechnoWeb: The community develops wide sets of innovations, ranging from product to service innovations with high impact on business activities, for instance the creation of new business models.

These are his struggles: He sees these viable examples of organizational innovation communities, in various industry sectors and in organizations of varying sizes, developing huge amounts of product and service innovations. Following the trend to establish communities within organizations, he introduced a wiki among employees to take advantage of employees' creative power and knowledge. This did not live up to expectations in his own organization as participation decreased drastically shortly after introduction. Consequently, he struggles with the question of whether there is a DNA that makes organizational innovation communities successful. Particularly, he asks himself if a Google-like culture (Hardy, 2005; Lashinky, 2006) or structures like that of 3M (Conceicao, Hamill, & Pinheiro, 2002) are needed as a pre-requisite for the success of innovation communities. In other words, he struggles with the question of *how organizational integration, by means of culture and structure, may influence organizational innovation communities and how he might create a supportive organizational environment.*

		**Tom** (top manager): 45 years old, studied business administration, worked as business consultant and plays golf
	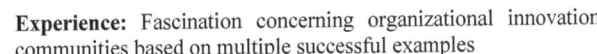	**Experience:** Fascination concerning organizational innovation communities based on multiple successful examples
		Struggle: Confusion about his role to provide supportive organizational contexts for organizational innovation communities

Table 1: Tom[4]

Ina is around 40 years old and is an innovation manager who studied informatics at a German university. She started to work for the organization directly after finishing her master thesis. She has worked in various departments of the organization and has considerable experience in project management and creativity techniques. She is an innovative person best imagined with colorful scarves and the latest gadgets on her fingertips. During her studies she installed the Linux systems on her computer out of curiosity but moved back to Windows later on. However, her curiosity in communities never vanished: She still has log-ins for several communities and is checks the latest entries on a regular basis.

This is what she experiences: She is intrigued by (i) the diversity of members collaborating across departmental boundaries and (ii) thereby creating significant innovations. For instance, Ina remembers the first innovation jams at IBM which addressed employees from diverse professional backgrounds successfully (Bungart & Köhler, 2009, p. 358). In a similar vein, Swarovski aims at inviting *all* employees with heterogeneous backgrounds to develop outstanding innovations (Erler, Rieger, & Füller, 2009, pp. 394–396). Moreover, she sees how great innovations are created at Cisco's I-Zone, crossing traditional departmental or cultural borders in a multinational organization (Roschek, 2009, p. 386). These examples seem to be as good as it gets, however she also observes some shortcomings.

These are her struggles: Besides the creative potential of diverse community members and collaboration across boundaries she frequently stumbles across community members

---

[4] All visualizations are created and drawn by the author and Anna-Katharina Hintelmann. Any use of visual material included in this thesis needs permission of the author.

insulting their peers or withdrawing from communication after controversies. Aware of these difficulties and their possible negative aftermath in the organization at large, she struggles with the question of how to take advantage of the inherent innovative potentials of organizational innovation communities while at the same time reducing the risk of conflict. In other words, she wants to know *how innovation development in organizational innovation communities unfolds and how this process may be facilitated.*

Ina (innovation manager): 40 years old, studied informatics, 15 years within organization, creative with the latest gadgets on her fingertips

Experience: Fascination concerning community-based innovation development at IBM and Cisco

Struggle: Confusion about her role to facilitate innovation development in organizational innovation communities

Table 2: Ina

Norm is around 30 years old. He works in the marketing department, having finished his training as a banker and joined the organization more than 10 years ago. Being active in several social networks like facebook, xing, and others is part of his job description. He lives in an idyllic town house together with his partner and their two-year-old girl. Norm identifies with the organization, does his work carefully and delivers his work on time. Recent financial crises and 'black Mondays' do not create feelings of fear because his organization manages to create constant revenues even in times of crisis.

This is what he experiences: Because he works in the marketing department he constantly gets feedback from customers about the company's products and services. Due to the length of his affiliation with the organization, he internalized strategic objectives long ago. Hence, he combines 'quasi-external' views on products and services while at the same time having strategic objectives of the organization in mind (Möslein & Neyer, 2009, p. 90). Consequently, besides increasing pressure to fulfill his daily tasks, he still thinks about innovative solutions and sometimes even submits a suggestion to the established idea management system. As such, he may best be described as a peripheral inside innovator, i.e. an employee whose job description does not include a responsibility to create innovations (Neyer et al., 2009).

These are his struggles: Even though Norm's submissions are frequently honored as being worthwhile to implement, the impact of his innovations is rather minor (Bansemir & Neyer, 2009). However, because he does not frequently develop innovations, he does not have the confidence to propose an innovation project that integrates his market and strategic insights. Moreover, he feels constrained by the mechanistic way in which innovations are submitted, i.e. by having to fill out a form. In his opinion, this is detrimental to the 'fun part' of innovating frequently associated with creative processes. So he asks himself how his innovation activities may be supported to unleash his full creativity and knowledge for innovation pursuits. In other words, his main question refers to *what influences his engagement in organizational innovation communities.*

	**Norm** ('normal' employee): 30 years, training as banker, more than 10 years within organization, thoughtful employee, active in facebook, xing, etc., and parent of one child
	Experience: Creation of innovations based on external feedback every time he is not washed away by daily tasks
	Struggle: Submission of innovation is not fun (filling out a form), limiting his engagement

Table 3: Norm

     Consequently, it becomes clear that **Tom, Ina** and **Norm** face particular challenges in implementing of innovation communities as a means to foster innovation development within their organization. They search for answers to their struggles to ensure that the application of organizational innovation communities substantially increases the effectiveness of innovation development. In other words, Tom, Ina and Norm aim to introduce a management innovation, i.e., to invent or implement "[…] a management practice, process, structure, or technique that is new to the state of the art and is intended to further organizational goals" (Birkinshaw, Hamel, & Mol, 2008, p. 825). This dissertation specifically sets out to give Tom, Ina and Norm some informed answers to their struggles.

     To enable them to implement organizational innovation communities as a management innovation three major questions are answered in the three empirical studies of this dissertation. First, Tom's question concerning *how organizational environments (or contexts) may influence organizational innovation communities and how he might create a supportive organizational environment* is tackled. Second, Ina's curiosity of *how innovation development in organizational innovation communities unfolds and how this process may be facilitated* is met. Third, Norm's question of *what influences his engagement in organizational innovation communities* is answered. By providing preliminary answers to these questions, this dissertation shines light on how organizational innovation communities can successfully be implemented as a management innovation. The structure of this thesis is provided in the following chapter.

2 Structure

The relevance and motivation for this thesis has already been stated: i.e. to help Tom, Ina and Norm to establish a management innovation by establishing an organizational innovation community. In a first step, Tom's, Ina's and Norm's experiences and struggles are checked against scholarly conversations in the field of community. Particularly, a systematic literature review categorizes and summarizes published articles, ultimately identifying research directions underpinning Tom's, Ina's and Norm's intuition (Part II). In a second step, three empirical studies shed light on the identified research directions. Tom's struggle concerning the antecedents of organizational contexts and the creation of supportive organizational contexts is addressed first (Part III). Subsequently, Ina's interest concerning the nature of innovation development and how to nurture it is met (Part IV). Lastly, Norm's struggle concerning a community member's impetus to engage in organizational innovation communities is tackled (Part V). In a third step, the findings of the empirical studies are reflected in terms of embedding the findings in context, summarizing key findings and displaying the interconnection between the parts (Part VI). In what follows, detailed information is given on each part.

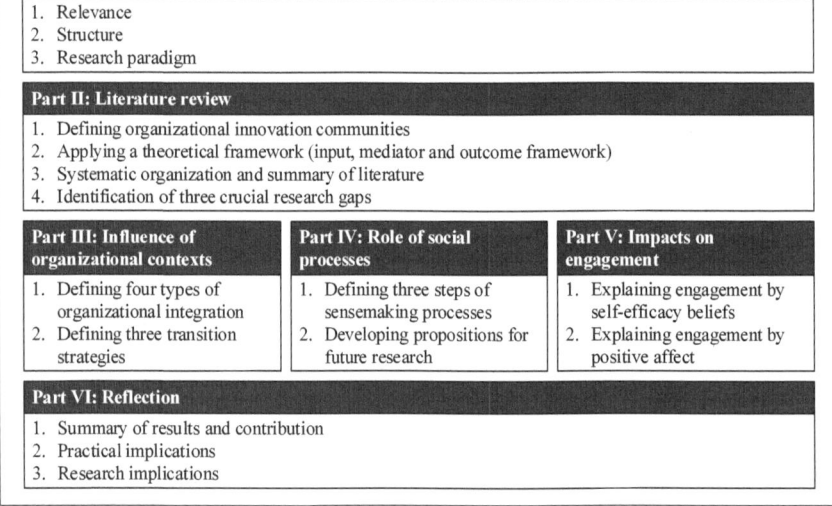

Part I: Introduction
1. Relevance
2. Structure
3. Research paradigm

Part II: Literature review
1. Defining organizational innovation communities
2. Applying a theoretical framework (input, mediator and outcome framework)
3. Systematic organization and summary of literature
4. Identification of three crucial research gaps

Part III: Influence of organizational contexts	**Part IV: Role of social processes**	**Part V: Impacts on engagement**
1. Defining four types of organizational integration 2. Defining three transition strategies	1. Defining three steps of sensemaking processes 2. Developing propositions for future research	1. Explaining engagement by self-efficacy beliefs 2. Explaining engagement by positive affect

Part VI: Reflection
1. Summary of results and contribution
2. Practical implications
3. Research implications

Figure 1: Structure of the thesis

The introductory **Part I** contains the basic impetus for this dissertation, presenting Tom's, Ina's and Norm's struggles of applying innovation communities as a means for innovation development within organizations. Furthermore, an outline of contents of this thesis is provided along with detailed information concerning the context in which this research is embedded.

Part II presents research in the field of communities that has been conducted over the past 12 years. Tom, Ina and Norm profit from this part as they are able to update their knowledge on communities and check their intuition against scholarly research conducted in this area. For this, it discusses the nature of communities, identifies differences and similarities underlying various types of communities, and derives a definition of organizational innovation communities. The review of community literature systematizes representative studies published by applying an input, mediator, and outcome framework. Altogether, 18 distinct factors are found that comprehensively describe relevant contributions in community literature in a condensed form. Distinct sets of avenues for future research are provided. Juxtaposing these avenues for future research and Tom's, Ina's and Norm's struggles, three major gaps in research concerning organizational innovation communities are identified: 1) exploring the influence of organizational contexts, 2) exploring collaborative social processes and 3) exploring influences on employee's engagement.

The first empirical part (**Part III**) addresses Tom's major struggle and the first gap of current research. It tackles the questions of how organizational contexts influence organizational innovation communities and how supportive organizational contexts may be shaped. Structuration theory serves as theoretical foundation to attain this aim. This empirical study builds on 12 in-depth case studies spanning various industry sectors – such as ICT, financial services, and engineering – and ranging from small to large-sized organizations. To help close the above-mentioned gap, the influence of organizational contexts on innovation activities and outcomes are explored. Four types of organizational innovation communities are found, characterized by distinct sets of innovation activities and outcomes resulting from varying degrees of cultural and structural integration. Moreover, the data analysis reveals three major transition strategies to alter organizational integration, whether cultural or structural. These are the initiation strategy, negotiation strategy, and narration strategy. Hence, Tom substantially profits from this exploration as he learns about the *influences* of culture and structure on organizational innovation communities and *strategies* to shape these organizational variables.

The second empirical part (**Part IV**) backs Ina's major struggle and tackles a lack of in-depth understanding concerning collaborative and social processes for innovation development in organizational innovation communities. It refers to the second major gap in scholarly work. Sensemaking theory is applied as a theoretical lens to explore these social processes for innovation development. Applying an action research approach in three organizations over a time-period of up to four years, data analysis builds on various data sources, including observations, interviews, attended meetings and presentations, and other data sources. However, the focal point of data analysis refers to observations of community members collaboratively developing innovations via a community platform. Results provide in-depth understandings of the social process of innovation development in organizational innovation communities. Specifically, the data analysis shows that a distinct form of cohesion, i.e., nested cohesion, helps to bind weakly connected community members together. Moreover, 'truly' collaborative innovation development on the community platform ignites a special form of flow, i.e., collaborative flow that is shared by multiple community members. Lastly, community members possess the abilities to integrate various perspectives through collaborative transformation. These findings drastically increase efficiency and effectiveness of innovation development in such communities. Ina learns about how to unleash the creative potential in collaborative innovation development in organizational innovation communities.

The last empirical study (**Part V**) extends previous research exploring employees' pro-active participation in the innovation activities of organizational innovation communities. Thereby, the influence of cognitive and affective states on knowledge exchange in such communities is examined. Using two experimental pre-test–post-test experiments, the data analysis reveals that both induction of self-efficacy and positive affect increase knowledge exchange, which is a pre-condition for innovation to occur. Importantly, the studies suggest that the induction of self-efficacy leads to a higher extent of knowledge exchange than the induction of positive affect. Norm learns from this part why community members engage in organizational innovation communities.

Part VI summarizes key contents of this thesis and describes contributions. Moreover, implications for practice and avenues for future research are provided. In sum, Tom, Ina and Norm update their knowledge on communities (Part II) and solve Tom's (Part III), Ina's (Part IV), and Norm's struggles (Part V). Part VI offers reflections on the results of this thesis. The following figure displays the structure of the thesis.

Concerning the presentation of contents, this thesis aims at fulfilling somewhat contradictory objectives: (i) displaying results in an easily understandable manner and (ii) providing in-depth insights which meet the standards expected from a research project. To achieve this challenging aim, **visualizations** of key content are frequently provided. These visualizations summarize the content described in the text in a narrative manner and offer a tool for navigating through the book in a more intuitive way. Readers should be able to follow the contents of this thesis by reading the visualizations. It is hoped that the reader may find it interesting to follow both the visualized and scientific content.

3 Research paradigm

Tom, Ina and Norm may be especially interested in the methodology of this thesis as an indicator of whether the results will withstand the 'practice-test'. In short, this thesis follows Witte's call for research in business administration to derive scientific results **and** practical implications at the same time (Witte, 1972, p. 7). To achieve this aim, the overarching research paradigm may best be described as a design science paradigm.

Design science may be defined as a problem-solving paradigm to make efficient and effective use of innovative information systems (Hevner, March, Park, & Ram, 2004; Denning, 1997; Tsichritzis, 1998). Particularly, design science creates and evaluates information system artifacts in social contexts (Hevner et al., 2004, p. 77). Hence, focal points of analysis refer to artifacts created as well as identification of useful ways to apply them. In sum, design science is a valuable paradigm to study phenomena in which information systems play a crucial role for valid scientific results as well as practical implications.

Particularly, Hevner, March & Park (Hevner et al., 2004, p. 81) emphasize that design science is particularly valuable to study **wicked problems**, i.e. challenges concerning the interplay of information systems and (i) constraining contexts, (ii) complex interactions and (iii) individual influences. Related to the first point, Tom may profit from a design science approach as he wants to design the organizational context for organizational innovation communities. In reference to the second point, the study of the complexity of interactions to develop innovations in organizational innovation communities, with which Ina is struggling, also benefits from a design science approach. Lastly, Norm also gets valuable insights as individual influences, such as motivation, are explicitly part of research agendas in design science. Taken together, the research paradigm of design science is a reasonable point of departure for studying organizational innovation communities.

To tackle these difficult problems, **guidelines** are offered (Hevner et al., 2004, pp. 82–95). The following guidelines are worth mentioning at this point to describe the background of the thesis: 1) design as an artifact, 2) design evaluation, and 3) design as a search process. First, *design as an artifact* describes "[...] a purposeful IT artifact created to address an important organizational problem" (Hevner et al., 2004, p. 82). To address the important question of how to transfer the trend of innovation communities into the organizational context an IT artifact is created in the context of the project 'Open-I: Open Innovation within the Firm' (funded by the German Federal Ministry of Education and Research and the European Social Fund).[5] The project may best be understood as an interdisciplinary and longitudinal study. A core team of researchers – including six professors in German universities, four post-doctoral researchers, eight PhD candidates,[6] and 12 part-time student researchers collaborate interdisciplinary. The research consortium includes computer scientists, information systems scientists along with business economists and psychologists. Furthermore, a core team of practitioners, including three CEOs of three large German companies, five additional board members in these companies, eleven innovation managers along with several trainees, accompany the project over a period of up to four years. The Open-I platform, as the IT artifact, is designed applying a four-step approach. In a first step and in the beginning of the project, literature is consulted to identify general aspects that have to be taken into account when designing community platforms at large or innovation communities more specifically. In a second step, interviews are conducted to find additional aspects ensuring the quality of the platform. In a third step, prototypes are developed to

---

[5] FKZ: 01FM07053, 01FM07054 and 01FM07055.
[6] Three of these eight PhD candidates hold a PhD degree at the time of writing.

include relevant aspects as well as to stimulate discussion of unenclosed relevant aspects. These first prototypes include Powerpoint visualizations as well as functional Excel sheets. Whereas the Powerpoint visualizations are used to ensure usability even before programming, Excel sheets simulate the key functions and interrelations of the sites of the platform. In a fourth step, the Open-I platform has been programmed adapting an open source solution provided by Elgg and PHP. In sum, knowledge from various disciplines has influenced the design of the Open-I platform (Bansemir, Habicht, Neyer, & Möslein, 2011). The following screenshot visualizes the starting site of the Open-I platform (Reinhardt, Wiener, Frieß, Groh, & Amberg, 2012).[7]

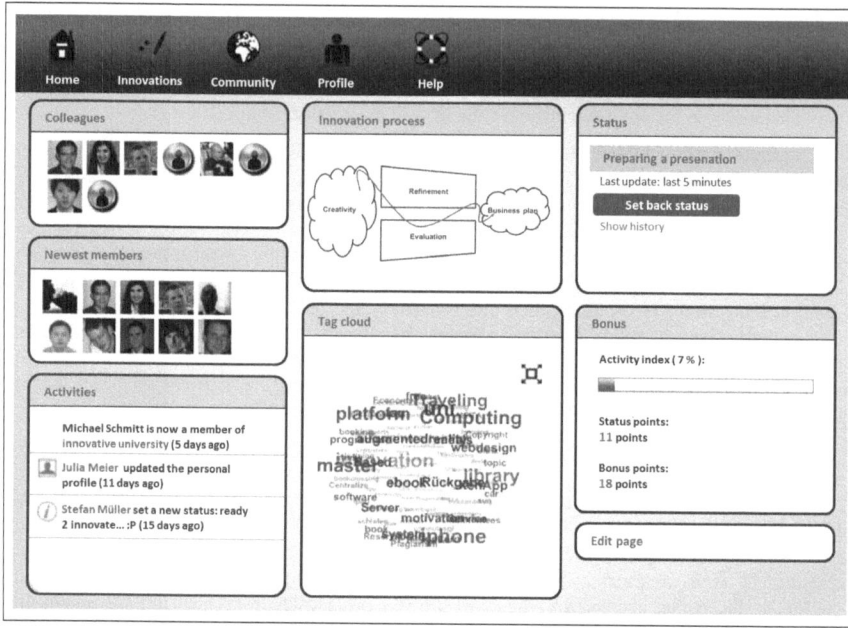

Figure 2: The starting page of the Open-I platform

Second, *design evaluation* is defined as the rigorous assessment of the quality and usefulness of the designed IT artifacts (Hevner et al., 2004, pp. 82–95). Evaluation of the Open-I platform is executed iteratively throughout the project. Altogether, three major evaluation steps are conducted. In a first step, the designed prototypes are evaluated. For this, the core research team intensively discusses the appropriateness of the platform. In a second step, the programmed Open-I platform is critically assessed by student researchers who actually collaborate on the platform. After some iterations of this assessment, the platform is provided to the team of practitioners in a third step. Again iterative evaluation and refinement is applied to further the performance of the IT artifact: the Open-I platform. Additionally, to derive trustworthy results more formalized evaluations of the artifact are conducted. For instance, interviews are conducted as well as experimental studies to assess specific functions

---

[7] The most important sites of the platform are provided in Annex A. The platform is accessible at www.dev.open-i.org.

of the platform (Reinhardt et al., 2012; Frieß, Groh, Reinhardt, Forster, & Schlichter, 2012; Schwarz & Bodendorf, 2012).

Third, the aim of *design as a search process* is "[...] to discover an effective solution to a problem [...] that can be implemented in the business environment" (Hevner et al., 2004, p. 88). It is crucial to mention that 'solutions' not only encompass the technical artifact but also the contexts, social processes, etc., as these factors ensure the appropriate use of artifacts. This approach has been an inherent part of the Open-I project since it emphasizes the crucial interrelation of the technical and the social. During the project, information and knowledge is constantly exchanged by means of regular trustworthy interactions. This means that researchers naturally visit companies' sites as part of their job. Particularly, in 'hot' phases of the project weekly visits are common, complementing intensive telephone or email communication between researchers and practitioners. Besides the provision of needed technical infrastructure, practitioners learn about how to successfully apply organizational innovation communities. In sum, scientific results are fed back to the companies involved and applied in real settings.

In conclusion, this thesis is built on a huge collaborative effort – between researchers of various backgrounds and practitioners in three German organizations – to transfer the trend of innovation communities to organizational contexts and thus to establish a management innovation. Applying a design science paradigm increases the validity of empirical findings because they are developed in real-life settings. Moreover, the applicability of research is fostered as managers may be more likely to transfer the resulting knowledge to their work contexts.

Part II Systematic literature review

1 Setting the stage[8]

The following sections provide useful background information for Tom, Ina, and Norm to update their knowledge and check their intuition against scholarly publications. First, the history of research investigations and writings are briefly outlined and constraints of this stream of literature are displayed. Moreover, a definition of organizational innovation communities is provided, juxtaposing various definitions discussed in literature. Lastly, a framework for the analysis of scholarly works is presented.

1.1 A brief history and constraints of community literature

It is important to mention that community research looks back to a long tradition of research attention in academic writings. Historically, examples include Durkheim's (1933) discussion of communities in comparison with solidarity or Etzioni's (1968) definition of collectives as normative based relationships. Whereas these works may be seen as sociological or philosophical works, Lave & Wenger's (1991) and Brown & Duguid's (1991) notions of 'legitimate peripheral participation' and 'community of practice' initiated the emergence of an extensive stream of literature on organizational communities within the domain of business administration. For instance, Wenger's (1998) book on communities of practice led to an explosion of academic writings (Amin & Roberts, 2008, p. 355). The book, with more than 15,000 citations, as of 2011 (Harzing, 2011), has proven to be one of the most influential works in organizational community literature and business administration in general.

In the last decade communities that cross organizational boundaries (i.e., boundary-less communities) have gained considerable attention mainly due to the achievements of open source software development such as Linux twenty years ago. Individuals working in virtual environments, mainly interacting on community platforms, characterize this form of community. For instance, Lerner & Tirole's (2002) article on open source communities led to a dramatic increase of scholarly work (with more than 1,800 citations, as of 2011, see Harzing, 2011). However, even though community research is a well-established phenomenon in academic writing (Adler, 2001), it tackles three major challenges:

First, one finds *inconsistent definitions* of community concepts (Probst & Borzillo, 2008; Mørk, Aanestad, Hanseth, & Grisot, 2008, p. 14; Gerybadze, 2007, p. 202), ranging from communities of practice (Wenger, 2000) to strategic communities (Kodama, 2006, p. 50). Thereby, these streams of literature do not differentiate between findings that are applicable to communities in general and those that only affect a specific type of community. Second, there is a lack of *sound and commonly accepted theoretical foundation*. Besides the fact that a decent number of theoretical articles on communities are available (Roberts, 2006; Brown & Duguid, 1991; Etzioni & Etzioni, 1999), a resilient framework is still missing, mainly due to the incomprehensive reflection of widespread academic works (Roberts, 2006; Wenger, 2000; Amin & Roberts, 2008; Brown & Duguid, 2001; Lynn, Mohan Reddy, & Aram, 1996). Third, the findings of studies on communities often *remain unrelated*. To advance research in the field it seems necessary to connect studies in a more sophisticated way (Kimble & Bourdon, 2008; Venters & Wood, 2007; Tarmizi & de Vreede, 2005).

---

[8] Part II is based on a previous conference article (Bansemir, Neyer, & Möslein, 2009) presented and discussed at the 2009 annual conference of the European Academy of Management (EURAM) and an unpublished working paper (Bansemir, Neyer, & Möslein, 2010) at the 2010 research colloquium Innovation and Value Creation in Beilngries.

To deal with these challenges, especially concerning the improvement of theoretical foundations and the fragmentation of empirical studies, a systematic literature review is conducted. The literature review aims at 1) defining organizational innovation communities, 2) delivering a theoretical framework, 3) systematically organizing and summarizing the main challenges in a comprehensive but condensed form, and 4) presenting paths for future research.

With these objectives in mind, the following literature review first systematically identifies and examines 247 scholarly publications on communities. Second, the identified and examined publications are classified into six major categories. Based on this classification 18 factors are identified. These factors and their interrelations and links provide a map of emerging avenues for future research. After juxtaposing these avenues for future research, an agenda motivating the three empirical studies of this dissertation is presented.

1.2   Defining organizational innovation communities

The widespread application of the notion 'community' leads to diverse definitions in business administration. For the following review of existing literature it is important to understand that community literature splits into two sub-streams, i.e. organizational communities and boundary-less communities. Nine specific forms of communities have gained considerable attention: 1) communities of practice, 2) communities of interest, 3) innovation communities, 4) knowledge communities, 5) learning communities, 6) strategic communities, 7) online communities, 8) virtual communities, and 9) open source communities. In the following, similarities and differences are discussed.

First and most prominent, the notion of *communities of practice* refers to a shared practice that determines community activities, i.e. enhancing capabilities in a specific practice (Wenger, 2000; Amin & Roberts, 2008, p. 354; Brown & Duguid, 1991). Second, in a *community of interest*, community activities emerge around a common concern, which most often crosses different practices (Hussler & Rondé, 2007, p. 289; Fischer, 2001b, p. 4; Wenger, 1998). Compared to communities of interest, communities of practice exert higher levels of shared understanding as a consequence of shared practices (Fischer, 2001b, p. 4). Third, *innovation communities* intend to create innovation, ranging from early stages of ideation to implementation (Fichter, 2006a, p. 96; Lynn, Aram, & Mohan Reddy, 1997; van Oost, Verhaegh, & Oudshoorn, 2009; Hippel, 2005, p. 96). The notion of innovation community is not mutually exclusive to the above-mentioned notions of communities, as the innovation focus may be part of a practice (e.g., in R&D departments) or interest (e.g., in hobbyist communities Chesbrough, 2003). Fourth, *knowledge communities* refer to the articulation, collective reconciliation and application of knowledge as the main objective of community activities (Lindkvist, 2005, pp. 1195–1197; Lee & Williams, 2007). Lindkvist (2005, p. 1195) argues that knowledge communities may be seen as communities of practice that operate on knowledge. Fifth, *learning communities* center around members, acquiring a new repertoire of knowledge or behavior while engaging in collective learning in context (Plaskoff, 2003, p. 170; Brown & Duguid, 1991, pp. 47–50). Sixth, *strategic communities* refer to community members, especially from middle management, formulating strategies in ambiguous environments, in which predictions about future developments are difficult (Kodama, 2006, p. 50). Seventh, *online communities* refer to internet-based communities concerning practices, interests, etc. (Armstrong & Hagel III, 1996; Ebner, Leimeister, & Krcmar, 2009, p. 345). Eighth, *virtual communities* are characterized by their often dispersed nature which is counteracted by wide application of ICT, such as telephone, internet and video-conferencing. This form of community mostly refers to communities in the internet but also includes communities within organizations (Bieber et al., 2002; Ardichvili, Page, & Wentling, 2003). Ninth, *open source communities* mainly refer to communities in which main activities take place in the internet and center around the development of new software (Lerner & Tirole, 2002). Besides the various definitions of communities, it is important to mention that the first six types of communities mainly relate to communities within organizations whereas the latter three mainly refer to communities that cross organizations, i.e., boundary-less communities. A study of cross-citations shows that these two streams of literature are rather unconnected and emerge mostly independently.

Despite these diverse definitions, all types of communities share similarities. Based on Wenger (1998), four major characteristics of communities are identified. These are, 1) mutual engagement, 2) shared repertoire, and 3) shared objective (Amin & Roberts, 2008; Barab et al., 2003; Bogenrieder & Nooteboom, 2004; Lindkvist, 2005; Plaskoff, 2003; Wenger et al., 2002; Wenger, 2000). First, *mutual engagement* refers to reciprocal activities of community members, such as working together and actualizing each other with latest information and stories (compiled from Wenger, 1998, pp. 73–77; Wenger, 2000, p. 227; Amin & Roberts, 2008, pp. 354–355). Second, *shared repertoire* is defined as artifacts in the form of stories and symbols, as well as processes, rules, etc. (Wenger, 1998, pp. 82–84). Third, *shared objective* is best described as a community's atmosphere, mostly concerning shared goals (Wenger, 1998, pp. 77–82; Probst & Borzillo, 2008, pp. 342, 339; Thompson, 2005, p. 159). However, mutual engagement, shared repertoire and shared objectives are not given *a priori*, but evolve in course of *community activities* (Roberts, 2006, pp. 624–625). Hence, these characteristics of communities ignore important dynamics of communities.

Community activities are expressed in particular by 1) coherence, 2) shared creation and 3) negotiation. First, *coherence* refers to individual objectives that overlap with community activities (Fang & Neufeld, 2009, p. 10; Roberts, Hann, & Slaughter, 2006, pp. 984–985). Thus, individuals who expect benefits from community participation will be motivated to join (Jian & Jeffres, 2006, pp. 242–243; von Krogh, Spaeth, & Lakhani, 2003, p. 1235; Komito, 1998). In contrast, without coherence, individuals are not willing to engage in community activities. Second, the *creation* of shared repertoire depends on collaborative efforts, in which individuals mutually combine perspectives and worldviews in recurring cycles (Østerlund & Carlile, 2005, p. 98; Nonaka & Toyama, 2003, p. 5; Sawhney & Prandelli, 2000, p. 46; Bechky, 2003, p. 322). Third, *negotiation* refers to the collective establishment of objectives. In this process, objectives are constantly discussed, refined and communicated through social interactions among community members (Drath & Palus, 1994; Palincsar, Magnusson, Marano, Ford, & Brown, 1998, p. 17; Swan, Scarbrough, & Robertson, 2002, p. 492; Linehan & McCarthy, 2001, p. 146). By juxtaposing the above discussed differences and similarities, organizational innovation communities are defined by the following characteristics:

(1) *mutual engagement*, which frequently emerges from *congruence of interests*,

(2) *shared repertoire*, which is continuously and *discursively created*,

(3) *shared objectives*, which regularly result from *negotiation*, and

(4) *development of innovation*, which is the main raison d'être.

1.3   Towards a community framework: Input, mediator and outcome

Given the fragmented nature of community literature the application of an input, mediator and outcome framework is especially valuable to guide a systematic organization of derived factors. Tom, Ina and Norm may find this framework especially useful, as they are able to search for topics in an organized and convenient manner. In this way, input factors are modeled in a multilevel nature in which individuals are nested in communities, which are themselves nested in a broader organizational context (Cohen & Bailey, 1997). Mediators consist of processes and emergent states. Processes describe interactions of members and therefore the transformation of input factors into outcomes. Emergent states do not relate to processes, but to conditions, e.g., states of mind resulting from input factors (Marks, Mathieu, & Zaccaro, 2001). Outcomes refer to products as results of inputs and mediators. They can be described in qualitative and quantitative terms (Mathieu et al., 2008, p. 412). Additionally, as communities function in a cyclic manner, the process from input to outcome is repeated in feedback loops (Ilgen, Hollenbeck, Johnson, & Jundt, 2005). The following figure displays the input, mediator, outcome framework explained above.

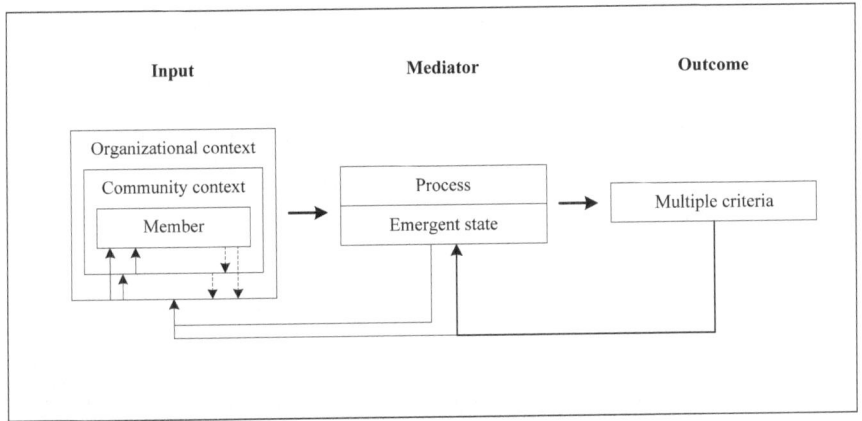

Figure 3: Input, mediator, outcome framework (based on Mathieu et al., 2008a, p. 413)

2 Organizational communities: A literature review

In the following literature review, the main focus is organizational communities. In other words, communities that cross organizational boundaries, such as online communities, open source communities, and virtual communities, are (for the most part) excluded from the presentation of contents in this chapter. Because findings related to boundary-less communities are relevant, especially for the design of the IT artifact, a summary of contents is provided in Annex B and a discussion of this stream of literature is provided at the end of this part. Key findings and gaps in research for these communities are summarized at the end of this part and contribute to the motivation to conduct three empirical studies. Subsequently, the method applied for analyzing organizational community literature is presented, along with a summary of key insights.

2.1 Research methods

To ensure coverage of relevant literature a threestep research approach was applied. As shown above, literature on organizational communities lacks clear annotations (Plaskoff, 2003, p. 167). Therefore an initial literature screening was conducted to identify relevant search terms which fit the phenomenon under research. Search terms were identified by screening abstracts and key words of articles that cited Brown & Duguid's (1991) seminal article addressing communities of practice. This procedure was stopped as soon as the key words were redundant and no new search terms were added. Throughout the subsequent process of reviewing the literature, search terms were added and articles supplemented.

In the next step, the database EBSCOhost Business Source Complete is used to conduct the actual search. Amin & Roberts (2008, p. 355) show that the number of publications rapidly increases in the period from 1998 to 2005, starting from a low level. Therefore the search was restricted to the last twelve years (starting with 1999, ending with 2011) to increase the percentage of relevant articles. As a first result, a list with around five hundred publications was generated. In addition, nearly thirty top management journals[9] were screened, issue by issue, to reduce the risk of missing relevant articles that did not include the search terms. Twelve additional articles, which were not identified with the key word search, were added.

Next, articles were independently identified by two reviewers, by screening at least title and abstract. Results were compared and in cases of disagreement, articles were read by both reviewers and discussed. The identified articles were read, short summaries were created, and dependent and independent variables listed. In addition, the articles were tagged with two to four labels. Based on the summaries, variables and labels, distinct input, mediator and outcome factors have been independently identified by two researchers and compared. All articles are assigned to one factor. The overall goal of this clustering is to derive factors that

---

[9] The following top management journals were additionally screened: *Journal of Management Information Systems, Administrative Science Quarterly, Academy of Management Journal, Academy of Management Review, Organization Science, Journal of Management, Strategic Management Journal, Journal of Business Venturing, Research Policy, Journal of Product Innovation Management, Organizational Behavior and Human Decision Processes, Creativity and Innovation Management, Organization Dynamics, Organisation Studies, Journal of Management Studies, Strategic Organization, Management Science, Entrepreneurship Theory and Practice, Sloan Management Review, Communications of the ACM, R&D Management, International Journal of Technology Management, IEEE Transactions on Engineering Management, Small Business Economics, Technological Forecasting and Social Change,* and *Journal of Small Business Management.*

are characterized by internal homogeneity but external heterogeneity. This step is challenging because most publications do not address only a single factor. In all cases, articles were summarized under the best fitting factor. Based on an input, mediator, outcome framework (see previous chapter) with its six major categories, 18 detailed factors were identified and explained in detail. In the end, a total of 131 articles (within the literature of organizational communities), published in 49 different academic journals, three collected editions and four monographs, were taken into account.[10]

2.2 Input factors

The following input factors, i.e. the main factors that influence outcomes through emergent states and processes, were put under the microscope for closer examination: Organizational context, community context, and members. Tom (the top manager) is probably most interested in the organizational context, since there he can find information that may help him resolve his struggles (i.e. how do organizational contexts influence organizational innovation communities and how can these contexts be fostered?). Ina profits most from studying the community context as she finds information about how social processes for innovation development unfold. Norm learns about individual factors that influence his community activities and outcomes from insights on the member level. Despite the fact that Tom, Ina and Norm are mainly interested in resolving their specific struggles, learning about all input factors may be interesting for them to develop a better understanding of the topic from researchers' perspectives.

2.2.1 Organizational context

First, I try to satisfy Tom's need for information about how organizational contexts influence mediators and outcomes. Several organizational input factors and their influence on mediators and outcomes are discussed within organizational community literature. The following section summarizes those factors that receive substantial consideration: strategy, structure, culture, and ICT. The following table visualizes and summarizes key aspects. Additionally at the end of this section a table displays detailed information about articles published at the organizational level. The table includes information about the main focus of the study, links to other factors as they were mentioned in the articles, methods applied and the type of community to which the article originally aimed at contributing (e.g., communities of practice, knowledge communities, etc.).

Tom probably supports the notion that organizational communities have to serve strategic objectives as agreed upon by most researchers in the field. **Strategy** in organizational community literature is described as the purposeful exploitation of communities to achieve defined strategic objectives (Kodama, 2005b, p. 896). Kodama (2005b, p. 896) emphasizes the strategic role of communities to "[...] form and implement concrete business concepts and ideas." It includes two sub-topics, i.e., environment and capabilities (Lynn et al., 1996, p. 103; Kodama, 2005b, p. 896).

---

[10] Articles included in the literature review are made publicly available by the author in the group 'organizational innovation communities' in Mendeley (www.mendeley.com).

Strategy refers to the strategic role [right] organizational communities fulfill to achieve organizational objectives. Organizational communities are sensitive to changes in the organization's *environment* [upper left] and also deliver needed *capabilities* [lower left] to adapt to environmental changes

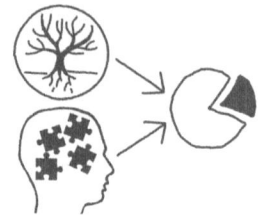

 Organizational communities should be purposefully exploited to serve *strategic objectives* of organizations, also to legitimate needed resources

 Structure [right] describes organizational factors in support for organizational communities, such as *resources* [upper left], *support by incentive scheme* [middle left] and *decentralized* [lower left] work organization

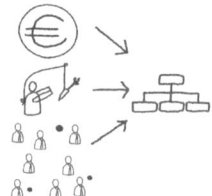

 Structure may be seen as needed requirements for organizational communities to flourish

 Culture refers to the supporting role of norms, values and beliefs [right] for organizational communities. *Legitimation*, i.e. acceptance that organizational communities contribute to strategic objectives [upper left], *awareness*, i.e. presence of communities' results [middle left], and *recognition*, i.e. appreciation of community activities, including top management [lower left], are major influences of supportive organizational cultures

 The organization's *culture* may support or hinder community activities, as too much cultural support stands against the democratic nature of communities

 ICT relates to electronic infrastructure that enables community activities. *Constant adaptation* [middle], i.e. inclusion of new technical features and termination of unsuccessful features, is a major driver for effective community activities

 ICT has to be constantly adapted to new features available, such as provided by Social Software and Web 2.0

Table 4: Organizational context: Strategy, structure, culture, and ICT

First, *environment* refers to the orientation of the organizational community towards environmental conditions in which the organization is embedded. Organizational communities need to reflect these conditions to generate strategic value (Lynn et al., 1997, p. 143; Pisano & Verganti, 2008, p. 86). Second, *capabilities* refer to unique sets of knowledge and experience community members may have, which potentially lead to superior problem-solving (Pisano & Verganti, 2008, p. 86; Lynn et al., 1996, p. 103). This is especially true if capabilities bridge between environmental challenges and strategic objectives.

Additionally organizational communities have to be supported by **structures**. For instance, Brailsford (2001, p. 20) and Lee & Choi (2003, p. 179) emphasize that structural variables on the organizational level play a crucial role to enable community processes, such as knowledge processes, and outcomes. Structures are frequently associated with antecedents, set up by the organization in support of organizational communities (Lee & Choi, 2003). These structures may be summarized under the terms resources, incentive scheme and decentralization (Venters & Wood, 2007, p. 360; Lee & Choi, 2003, p. 179; Brailsford, 2001, p. 25).

First, *resources* are needed to energize organizational communities, in the sense that they facilitate processes and ensure valuable outcomes. For instance, sponsors are seen as an important resource, as they invest time to support the community (these activities are explicitly part of the sponsor's job description) (Brailsford, 2001, p. 25; Tarmizi & de Vreede, 2005, p. 3545). Second, another major driver to increase individuals' willingness to contribute to organizational communities is the *incentive scheme* in place (Kimble & Bourdon, 2008, p. 466). Studies show that incentive schemes have a positive influence on processes such as knowledge exchange, because they incentivize employees' engagement in the community (Kimble & Bourdon, 2008, p. 466). Third, the overall design of organizations in a *decentralized* manner positively influences effective collaboration in organizational communities (Venters & Wood, 2007, p. 360; Lee & Choi, 2003, p. 179). It is shown that decentralized structures ask for similar working styles as in communities. Consequently, employees find it easier to work in communities if they can build on experience in decentralized organizational structures. In sum, structural factors represent basic requirements that should be available to organizational communities to flourish.

Culture is identified as a third major driver of organizational communities. For instance, Kimble & Bourdon (2008, p. 466) emphasize the supporting role of organizational culture to foster organizational communities. Community activities have to be an integral part of the organization's norms and values. Particularly, the organization's culture has to embrace employees' community activities. Three sub-topics characterize organizational culture as an input factor: Legitimation, awareness and recognition.

First, researchers agree that despite the positive effects of community-based activities, such as those reported in the study of copier mechanics by Orr (1986) and Brown & Duguid (1991), *legitimation* by the organization at large is often still an issue of major concern. For instance, the exchange of implicit knowledge in communities is often facilitated through narration – or, in other words, storytelling. Often this important mechanism is misunderstood as unimportant conversations, whereas in truth implicit knowledge is exchanged and complex problem-solving enabled. To support the legitimation of community efforts Brailsford (Brailsford, 2001, p. 25), for instance, finds that legitimation increases as the community exhibits a clearly defined purpose that community members are able to communicate. Second, *awareness* of community efforts is a major element of a supportive culture. Studies show that it is crucial that employees know about community activities and ways to engage in these (Brazelton & Gorry, 2003, p. 25). For instance, public presentations in the form of market stands in entrance halls of organizations are found to create awareness. However, one may think of multiple other ways to create awareness, for instance presence in the intranet or posters upon many more. Third, Chua (2006, pp. 127–128) identifies *recognition* as an important driver for organizational communities on the organizational level. Recognition may be understood as appreciation of community activities. One mechanism frequently suggested by literature refers to top management attention. Top management attention may be granted by means of inviting active community members to present their work or to display top contributors at popular places (similar to the "employee of the week" at McDonalds).

In times of blogs, wikis, and social networks like facebook and LinkedIn it seems reasonable that information and communication technologies (**ICT**) are an increasingly important part of community activities. ICT refers to the electronic infrastructure that enables access to and storage of information, the identity representation of community members, and communication. Features to realize these functions include search and retrieve, public archives of experts, virtual meeting rooms and the provision of databases (Krumsvik, 2005, p. 38; Etzioni & Etzioni, 1999, pp. 242–246; Schwen & Hara, 2003, p. 260). Besides a vivid discussion of how different features support community activities (e.g., Reinhardt et al.) constantly adapting to ICT is identified as most important.

Constant adaption refers to the adaptation of existing features, the inclusion of new features and the exclusion of antiquated features due to rapid technical development of ICT. Barrett et al. (2004, p. 9) emphasize that the landscape of available features is constantly changing and hence it is a constant challenge to balance the usage of existing features with the introduction of new features to facilitate community collaboration. For instance, recent studies show that new features – especially concerning recent developments often summarized under the label of Web 2.0 or social software – bear tremendous potential to boost community activities (Novak, 2007, p. 626; Barrett et al., 2004, p. 9; Hildreth, Kimble, Wright, & de Luminy, p. 12; McDermott, 1999b, p. 112). However, integrating these features in the logic of existing community platforms under the premise of usability remains challenging, especially if they are designed to facilitate community activities (Barrett et al., 2004, p. 8; Etzioni & Etzioni, 1999, p. 247; Schwen & Hara, 2003, p. 260; McDermott, 1999b, p. 116).

Main topic:	Related topics:	Applied method:	Focus:	Study:
Strategy	ICT, status	Qualitative case study	Strategic community	Kodama, 2005b
	ICT, stickiness, outcome	Conceptual, descriptive case study	Innovation community	Pisano & Verganti, 2008
	Proximity, social process, centrality, seeding patterns	Theoretical	Innovation community	Lynn et al., 1997
	Proximity, social process, centrality, seeding patterns	Theoretical	Innovation community	Lynn et al., 1996
Structure	Motivation, knowledge process, learning process, outcome	Descriptive case study	Knowledge community	Brailsford, 2001
	ICT, trust, knowledge process, learning process, social process	Quantitative questionnaire	Knowledge community	Lee & Choi, 2003
	Seeding pattern, status, knowledge process, social process	Case study	Community of practice	Tarmizi & de Vreede, 2005
	ICT, culture, strategy	Qualitative case study	Virtual community	Rosenbaum & Shachaf, 2010
	Boundary object, trust	Participative action research	Community of practice	Venters & Wood, 2007
Culture	ICT, social process, outcome	Participative action research	Knowledge community	Brazelton & Gorry, 2003
	ICT, structure, strategy, motivation	Qualitative interviews	Community of practice	Kimble & Bourdon,

			2008
Seeding patterns, structure, trust	Descriptive case study	Community of practice	Lank, Randell-Khan, Rosenbaum, & Tate, 2008
Structure	Descriptive case study / qualitative interview	Community of practice	Chua, 2006
ICT, structure, trust, boundary object	Participative action research	Community of practice	Dubé, Bourhis, & Jacob, 2005
ICT	Knowledge processes, objectives, motivation, trust	Qualitative case study	Community of practice	Pan & Leidner, 2003
	Motivation, centrality, culture	Qualitative case study (based on secondary data)	Community of practice	Schwen & Hara, 2003
	Cultural, centrality, objective, social process, boundary object	Participative action research	Community of practice	Krumsvik, 2005
	Social process, trust	Theoretical	Community of practice	Etzioni & Etzioni, 1999
	Objectives, knowledge process, learning process	Qualitative case studies (based on secondary data)	Knowledge community	Barrett et al., 2004
	Boundary object, knowledge process, social process	Quantitative and qualitative case study	Community of practice	Novak, 2007
	Trust, structure	Action research	Community of practice	MacDonald, 2008
	Strength of ties, proximity	Qualitative case study	Community of practice	Hildreth et al.
	Knowledge processes, objective, motivation, status, structure	Descriptive case study	Community of practice	McDermott, 1999b
	Learning process, knowledge process, trust	Conceptual	Community of practice	Lueg, 2000
	Knowledge process, proximity, stickiness, seeding pattern	Qualitative case study	Virtual community	Gammelgaard, 2010
	Outcome, boundary objects, ICT	Conceptual	Community of practice	Roberts et al., 2006

Table 5: Publications assigned to organizational context

Summarizing this section, the main focus concentrates on Tom's struggle to find information concerning the question of how to design organizational contexts that support organizational communities. He learns that strategy, structure, culture, and ICT all play crucial roles in facilitating organizational communities. Tom is astonished by the knowledge that research has already created. Some of the question marks have disappeared, yet some remain.

2.2.2   Community context

In the subsequent section Ina's confusion about how to facilitate community activities is addressed. Literature related to community contexts is consulted to clear up some of the fog in her mind. Particularly, literature on the community level provides Ina with two major means to foster community activities, i.e. boundary objects and seeding patterns. The table below summarizes and visualizes key findings from literature. Moreover, the table at the end of this section gives an overview of research done in this field, its focus, related topics, methods applied, and community focus.

Boundary objects [right] are defined as artifacts that help community members to bridge multiple intersecting social worlds. Three conditions are necessary for effective use of boundary objects: *openness*, i.e. undisguised mind sets of community members [upper left], *richness*, i.e. they have to capture divergent perspectives [middle left], and *explanatory*, i.e. they have to display the full complexity of an issue at hand [lower left]

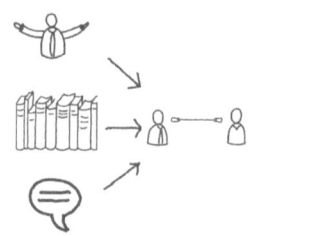

Boundary objects may have the potential to connect heterogeneous community members and capture the full complexity of the issues at hand

Seeding patterns [right] may be described as fertile seeding grounds for organizational communities to achieve objectives. Seeding patterns include *facilitators*, i.e. organizers coordinating community activities [middle upper left], *objectives*, i.e. goals directing engagement [upper left], *content feedback*, i.e. feedback concerning own contributions [lower middle left], and *performance feedback*, i.e. feedback concerning the community's performance [lower left]

Seeding patterns are a crucial factor in facilitating community activities. However, if hierarchy-related pressure is induced, community activities may suffer considerably

Table 6: Community context: Boundary objects and seeding patterns

Ina is still fascinated by the somewhat 'magic' mechanisms that make community members share understandings even though they often do not really know each other. One important aspect to make these 'magic' mechanisms work are, what most researchers call, **boundary objects**. They are defined as artifacts that perform a brokering role between multiple intersecting social worlds, e.g. community members who never meet, are integrated in different cultures, have different mother tongues, etc. (Star & Griesemer, 1989, p. 393; Star, 1989; Fischer, 2001b, p. 6; see also: Arias & Fischer, 2000; Fischer, 2001b, p. 6). Following Plaskoff (2003, p. 173) they may take the form of: a) common value systems, such as a common belief structures and future objectives (Gerybadze, 2007, p. 206; McDermott, 1999a, p. 4); b) accepted behaviors, such as ways of performing a task and shared language; and c) sense of belonging, including members' emotional feelings of relationship and belonging to a community (Joon Koh & Young-Gul Kim, 2003, p. 76). Examples of boundary objects include technical drawings among engineers, models of buildings among architects or source code for programmers, but also behavioral norms in flat shares or 'ways-of-doing-things' in organizations. However, boundary objects do not only facilitate the creation of shared understandings or knowledge processes, but also support social processes such as communication among community members (2003, p. 326). Three major sub-topics are frequently discussed in the organizational community literature: 1) openness, 2) richness, and 3) explanatory.

First, *openness* refers to undisguised mindsets of community members that enable effective use of boundary objects. Particularly, Garrety, Robertson, & Badham (2004, p. 357)

show that community members prioritizing individual interests and worldviews prohibit boundary objects to unfold their brokering role. Moreover, Bechky (2003, pp. 322–323) demonstrates how assembly workers and engineers use boundary objects while solving manufacturing problems. The example shows that boundary objects do not fulfill their role if engineers do not believe assembly workers in their expertise (and vice versa). However, as soon as they accept the expertise of their counterparts, shared understandings are instantly created around boundary objects. Consequently, community members exerting an open mindset are able to make use of boundary objects' potential to create shared understandings. Second, *rich* boundary objects, i.e. representing knowledge from heterogeneous community members, are able to explicate comprehensive understandings of complex issues (Bechky, 2003, p. 314). In other words, if community members manage to create boundary objects that include their divergent experiences, worldviews, professional or even cultural backgrounds, they develop a shared understanding of important aspects concerning complex issues. Reflecting observations of boundary object creation, Arias & Fischer (2000, pp. 6–7) emphasize that boundary objects are an explication of the combined knowledge of all community members. Third, boundary objects are *explanatory*, in the sense that they embody rich understandings, complex relations and may be of tacit nature. For instance, stories have the inherent power to illustrate otherwise hardly verbalizable interrelations. Another example of boundary objects that embody tacit knowledge concerns demonstrations and the repetition of procedures, actions, etc. (Bechky, 2003, p. 324).

Despite these helpful insights concerning boundary objects, their usage is not fully understood (Bechky, 2003, p. 326). For instance, Fischer (2001b, p. 4) emphasizes that the heterogeneity of community members has the potential to create rich understandings. In contrast, Gerybadze (2007, p. 206) puts forth that heterogeneous community members often exert reduced efficiency of task fulfillment. However, Ina finds herself yet again puzzled as she struggles with the question of how boundary objects may enable the inclusion of heterogeneous members and efficient task fulfillment at the same time.

Furthermore, research hints at a bundle of patterns that help to set up and energize organizational communities (Swan et al., 2002; Thompson, 2005, p. 162). These patterns are frequently described as **seeding patterns**. In contrast to hierarchical order and control mechanisms, they cultivate a fertile community environment in a non-prescriptive and indirect way to ensure the effectiveness of community activities. The following four seeding patterns are emphasized in analyzed studies: 1) facilitator, 2) objectives, 3) content feedback, and 4) performance feedback.

First, Wenger, McDermott, & Snyder (2002, p. 12) have long advocated liable *facilitators* to be of major importance for organizational communities to function effectively. Facilitators perform the task of organizing meetings (Breu & Hemingway, 2002, p. 149; Probst & Borzillo, 2008, p. 341), time keeping (Woodland, Szul, & Moore, 2007, p. 76; Pemberton, Mavin, & Stalker, 2007, p. 65) and structuring of content (Thompson, 2005, p. 163). Despite the individual position of power within the hierarchical structure, facilitators experience equal status within the community (Pemberton et al., 2007, pp. 67–68). In other words, one's official role within the organization neither qualifies nor disqualifies one for a role as a facilitator within organizational communities. This means that Tom, Ina, and Norm would have equal opportunities of becoming a facilitator. Second, Probst & Borzillo (2008, pp. 342, 339) find that in addition to a democratic structure, organizational communities need to have clear *objectives* to facilitate the serious and rigorous engagement of community members. Points of focus or central challenges serve as objectives in organizational communities (Thompson, 2005, p. 159). In the example of Bechky (2003, pp. 322–323), the main point of focus relates to solving manufacturing problems. Additionally, time pressure, for instance in the form of time-restricted competitions, can ignite community activities

(Thompson, 2005, p. 159). Third, Woodland, Szul, & Moore (2007, p. 75) argue that timely *content feedback* nurtures community processes. Probst & Borzillo (Probst & Borzillo, 2008, pp. 340–343) find that prompt feedback encourages the testing of new ideas, solving common problems, and incorporating fresh ideas. In line with Thompson (2005, pp. 157–158), Breu & Hemingway (2002, p. 150) and Woodland, Szul, & Moore (2007, p. 72), this feedback may be provided by regular face-to-face meetings, as well as chats, instant messaging, etc. Fourth, *performance feedback* is an important means to demonstrate the effectiveness of community activities (Probst & Borzillo, 2008, p. 339). As community activities are mostly hard to access in terms of traditional performance measures, studies put forth the importance to show community members how their activities contribute to the objectives of the organization and to justify community activities in front of top management and other employees (Breu & Hemingway, 2002, p. 149; Probst & Borzillo, 2008, p. 339). For instance, Lakhani et al.'s (2007) article displaying the effectiveness of scientific community problem solving received considerable attention among scholars, as the effectiveness of community activities are clearly displayed in verifiable quantified measures.

Seeding patterns especially help Ina to understand how she may foster community activities. However, seeding patterns in the form of facilitators, objectives, content, and performance feedback have to be carefully applied: Besides the energizing potential of seeding patterns (Breu & Hemingway, 2002, p. 149; Woodland et al., 2007, p. 76) they may also hamper community activities by inducing hierarchy-related pressure (Probst & Borzillo, 2008, p. 342; Thompson, 2005, p. 163). Hence, when applying seeding patterns Ina has to carefully balance the degree to which facilitators engage, objectives are defined and communicated, content feedback is provided and performance feedback is given.

The following table (next page) summarizes key findings from literature referring to community context.

Main topic:	Related topics:	Applied method:	Focus:	Study:
Boundary object	Learning process, knowledge process	Participative action research	Community of practice	Bechky, 2003
	ICT, knowledge process	Descriptive case study	Community of interest	Fischer, 2001a
	ICT, knowledge process	Descriptive case study)	Community of interest	Fischer, 2001b
	Social process, boundary objects, seeding patterns, motivation	Reflexive action research	Community of practice	Garrety et al., 2004
	Strength of ties, social process, proximity	Theoretic	Innovation community	Gerybadze, 2007
	Seeding patterns, motivation, social process, structure, ICT	Quantitative questionnaire	Virtual Community	Joon Koh & Young-Gul Kim, 2003
	Structure, ICT, culture,	Conceptual	Community of practice	McDermott, 1999a
	Centrality, knowledge process, social process	Theoretical	Community of practice	Plaskoff, 2003
	Moderator, strength of ties, proximity	Theoretical	Community of practice	Wenger, 2000
Seeding patterns	Seeding patterns, strength of ties, culture, outcome	Qualitative interviews	Community of practice	Probst & Borzillo, 2008
	Structure, motivation, strength of ties, knowledge process	Qualitative interviews	Innovation community	Breu & Hemingway, 2002
	Strength of ties, boundary object, knowledge process	Qualitative case study	Community of practice	Bettiol & Sedita, 2011
	Motivation	Participative action research	Community of practice	Schwen & Hara, 2003
	Seeding patterns, proximity, centrality, social process	Theoretical	Community of practice	Pemberton et al., 2007
	Boundary object, ICT	Participative action research	Community of practice	Thompson, 2005
	Culture, proximity, motivation, social process	Descriptive case study	Learning community	Woodland et al., 2007
	Social process, seeding patterns, proximity, outcome	Conceptual	Learning community	Gassmann & Enkel, 2006

Table 7: Publications assigned to community context

2.2.3   Member factors

In the following section the focus shifts from Ina to Norm. He is interested in what factors drive community members to actively participate in organizational innovation communities. Researchers may give him manifold answers to his question. In the following section the extensive body of literature around member factors is summarized. Most importantly, Norm will be interested that factors concerning motivation, trust, and status play a crucial role for community members to engage actively. The table below encapsulates key aspects of member factors while table 9, at the end of this section, displays corresponding publications, community focus, methods applied, and related topics.

Motivation [right] ignites goal-oriented behavior of community members. It may have three major sources: *utilitarian motives*, i.e. cost-benefit considerations [upper left], *normative motives*, i.e. internalized community culture such as norms [middle left], and *collaborative motives*, i.e. enjoyment of task itself [lower left]

 Motivation structures change in the course of community activities from initial utilitarian motives to normative and collaborative motives

 Trust [left] is characterized by community members' beliefs that they will not take advantage. Four types of trust may be distinguished: *dispositional trust*, i.e. a community member's general trusting attitude [upper left], *calculus-based trust*, i.e. reliance on a regulatory system of trust violation [upper middle left], *information-based trust*, i.e. third parties reporting trustworthiness [lower middle left], and *identification-based trust*, i.e. experience-based [lower left]

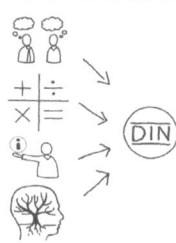

 Trusting attitudes also change over time and are dependent on experiences within organizational communities. Identification-based trust may only be achieved after several months or even years of intensive collaboration

 Status [right] refers to the social position of each community member within organizational communities. Status may be gained by *substantial contribution*, i.e. long-term and high quality involvement [upper left], *transformational communication*, i.e. informal, close and inspiring interactions [middle left], and *social brokerage*, i.e. circulation of new knowledge [lower left]

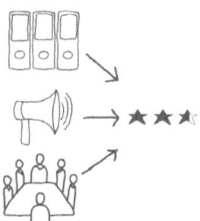

 Community members, who achieve *high status* gain considerable influence concerning future community objectives and coherence of the community

Table 8: Member factors: Motivation, trust and status

Norm's expectations that motivational aspects play a crucial role for explaining community members' engagement is supported by scholars in the field (Fang & Neufeld, 2009, p. 10; Roberts et al., 2006, pp. 984–985). **Motivation** is widely understood as the activation of goal-oriented behavior (Osterloh & Frey, 2000). Studies distinguish three major factors that potentially unleash motivation among community members (Jian & Jeffres, 2006, pp. 242–243; von Krogh et al., 2003, p. 1235; Komito, 1998): 1) utilitarian motives, 2) normative motives, and 3) collaboration motives.

First, *utilitarian motives* refer to individual's cost-benefit considerations (Ebner et al., 2009, p. 353; Fang & Neufeld, 2009, p. 11). In a nutshell, individuals driven by utilitarian motives engage in community activities if expected benefits outweigh expected costs, for instance in terms of time invested. Thereby utilitarian motives are associated with extrinsic motives that are based on rewards or incentives, such as financial rewards or attention (Ebner et al., 2009, p. 353). Second, *normative motives* drive members' activities based on internalized community culture (Fang & Neufeld, 2009, p. 11; de Valck et al., 2007, p. 252). In other words, if community members contribute on a regular basis because they are expected to do so, it is likely that they are driven by normative motives. Normative motives include community norms and beliefs, such as norms of reciprocity or trust (Fang & Neufeld, 2009, p. 15). Third, *collaborative motives* unfold if community members are driven by interaction and collaboration with other community members. Motivation occurs as an inherent of work contents, in the sense that community members enjoy the task they work on (Osterloh & Frey, 2000, p. 539; Fang & Neufeld, 2009, p. 11). Collaborative motives are therefore associated with intrinsic motives, as the work content is motivating for its own reason (Osterloh & Frey, 2000, p. 539; Jian & Jeffres, 2006, pp. 245–248).

Studies additionally show that motivation structures change over time. In brief, several studies show that utilitarian motives, driving initial engagement, change to normative or collaborative motives in course of engagement (Franke & Shah, 2003, pp. 172–173; Fang & Neufeld, 2009, p. 43; von Krogh et al., 2003, p. 1235; Bishop, 2007, p. 1891; de Valck et al., 2007, p. 253; Fang & Neufeld, 2009, p. 43). Reflecting research conducted on the motivational aspect, Norm's own experiences match research conducted. Still, it remains unclear why he seems to be motivated to engage while his peers are not to an equal degree. In other words, it remains uncertain how individual differences among community members are reflected in motivation structures (Bishop, 2007, p. 1891; Fang & Neufeld, 2009, p. 43).

Research emphasizes the superior role of **trust** for organizational communities, especially to enable effective knowledge and social processes (Adler, 2001; Chiu, Hsu, & Wang, 2006, p. 1877; Hsu, Ju, Yen, & Chang, 2007, p. 154; Roberts, 2006, p. 628; Nonaka, 1994). Trust is defined as implicit beliefs that other members behave with integrity, benevolence, and competence, and do not take advantage of situations (Chiu et al., 2006, p. 1877; Hsu et al., 2007, p. 154). Four aspects describe trust relationships among community members: 1) dispositional trust, 2) calculus-based trust, 3) information-based trust, and 4) identification-based trust (Soto, Vizcaíno et al., 2007, pp. 354–355; Hsu et al., 2007, pp. 157, 160).

First, *dispositional trust* describes underlying trusting attitudes of community members (Soto et al., 2007, p. 355; Chiu et al., 2006, p. 1877). For instance, community members who exert a high basic level of confidence in (unknown) community members may be described as a person with dispositional trust. Second, *calculus-based trust* refers to reliance on a system, including cost-benefit calculations in the case of violation of trust and possibilities of punishment (Hsu et al., 2007, p. 157; Soto et al., 2007, p. 355). Conversely, it is not based on properties of a relationship. For instance, buyers and sellers at ebay frequently exert calculus-based trust as they are confident that the system (ebay) moderates in case of trust violation. Third, *information-based trust* is based on assumptions that behavior outcomes are predictable as other community members report trustworthiness (Hsu et al., 2007, p. 157; Soto et al., 2007, p. 354). Information-based trust currently evolves in social interactions if someone says about a third person: "You can trust this person. I know her/him." Fourth, *identification-based trust* includes emotional bonds, mutual understanding and appreciation among community members which enable them to stand up for each other (Hsu et al., 2007, p. 160). This form of trust usually takes a lot of time to grow. Studies show that trust develops over time from calculus-based to identification-based trust if no reason is given to not trust a person (Hsu et al., 2007, pp. 165–166). However, little is known on how this development unfolds and how this development could be supported in organizational communities.

Lastly, the **status** given to community members is an important mechanism for organizational communities. Research shows that despite their chaotic appearance, organizational communities organize themselves to a large degree by community members with varying status levels (Fleming & Waguespack, 2007, p. 165; Lerner & Tirole, 2002; Hippel & von Krogh, 2003). Thereby, it is understood as a degree to which a community member is able to exert influence on the community in terms of legitimately creating coherence among community members, retaining the community's culture or providing objectives (Muller, 2006, p. 385; Fleming & Waguespack, 2007, p. 168). Status is given to community members if they exert the following behaviors: 1) substantial contribution, 2) transformational communication, and 3) social brokerage.

First, *substantial contribution* refers to input given by a community member that is characterized by exceptional quality or substantial quantity (Muller, 2006, p. 388; Sloman & Reynolds, 2003, p. 268; Wagenaar & Hulsebosch, 2008, p. 16; Kodama, 2001b, p. 84). Studies find that status is strongly related to contributions given by community members (Fleming & Waguespack, 2007, p. 178; O'Mahony & Ferraro, 2007, pp. 1092, 1099; Sloman & Reynolds, 2003, p. 268). For instance, Fleming & Waguespack demonstrate that the number of contributions of a specific community member determines the degree to which other community members follow his or her opinion. Second, *transformational communication* is characterized by informal, close and inspiring interactions across boundaries, such as professions (Fichter, 2009, p. 369; Fleming & Waguespack, 2007, p. 178). Fichter (2009) shows that if community members apply transformational communication to keep community activities thriving, they are likely to gain status. Third, *social brokerage* is understood as the translation of locally created meanings into accepted standards of the whole community by specific community members (Fleming & Waguespack, 2007, p. 170; Garavan, Carbery, & Murphy, 2007, pp. 41–42). In other words, locally created processes, solution, objectives, etc., develop into widely accepted behavior guidelines if a community member is able to generate broad agreement. Therefore social brokerage activities have the potential to bind community members together (O'Mahony, 2007, p. 1093; Sloman & Reynolds, 2003, p. 264). Members who are able to accomplish this task successfully gain status on return (Garavan et al., 2007, p. 36; Fleming & Waguespack, 2007, p. 178; O'Mahony & Ferraro, 2007, pp. 1092, 1099).

Individuals that exhibit abilities to contribute substantially, communicate transformational, and exert social brokerage are likely to gain high levels of status. On the other side of the coin, these members have the power to influence the community's course of action (O'Mahony & Ferraro, 2007, pp. 1079–1080; Garavan et al., 2007, p. 46).

Main topic:	Related topics:	Applied method:	Focus	Study:
Motivation	Knowledge process, trust, strength of ties, proximity	Qualitative case study	Community of practice	Ardichvili et al., 2003
	Objective, feedback, centrality	Conceptual	Community of practice	Bishop, 2007
	Seeding patterns, status, trust, structure	Quantitative case study	Innovation community	Franke & Shah, 2003
	Seeding patterns, outcome	Quantitative questionnaire	Community of interest	de Valck et al., 2007
	ICT	Quantitative questionnaire	Community of practice	Jian & Jeffres, 2006
	ICT, centrality	Qualitative and quantitative case study	Innovation Community	von Krogh et al., 2003
	ICT, seeding patterns, seeding patterns, objective, structure	Quantitative case study	Innovation community	Ebner et al., 2009

	Knowledge process, outcome	Theoretical	Knowledge community	Osterloh & Frey, 2000
	Outcome	Qualitative case study	Innovation Community	Fang & Neufeld, 2009
	Motivation, status, outcome	Quantitative questionnaire	Innovation community	Roberts et al., 2006
	ICT, culture, proximity, status, boundary objects	Theoretical	n.a.	Komito, 1998
Trust	Social process, culture, status	Theoretical	n.a.	Adler, 2001
	Strength of ties, objects, motivation, knowledge process	Quantitative questionnaire	Virtual Community	Chiu et al., 2006
	Structure, motivation, centrality	Conceptual	Community of practice	Roberts, 2006
	Motivation, knowledge process	Quantitative questionnaire	Virtual communities	Hsu et al., 2007
	Motivation, ICT	Conceptual	Community of practice	Soto et al., 2007
	Motivation, context, knowledge process	Quantitative questionnaire	Knowledge community	Szulanski, Cappetta, & Jensen, 2004
	Seeding patterns, seeding patterns, learning process, outcome	Qualitative interviews / quantitative questionnaire	Learning community	Edmondson, 1999
	Knowledge process, motivation, culture	Quantitative experiment	Knowledge community	Quigley, et al., 2007
Status	Learning process, social process, centrality, trust	Theoretical, proposition generating	Community of practice	Driver, 2002
	Trust, reputation, strength of ties	Quantitative case study	Innovation community	Fleming & Waguespack, 2007
	Motivation	Qualitative case study	Innovation community	Fichter, 2009
	Objectives, trust, reputation	Simulation	Community of practice	Muller, 2006
	Trust, social process, objectives	Participative case study	Community of practice	Garavan et al., 2007
	Centrality, social process	Qualitative / quantitative case study	Innovation community	O'Mahony & Ferraro, 2007
	Knowledge process, centrality	Descriptive case study	Learning community	Sloman & Reynolds, 2003
	Knowledge process, learning process, centrality, strength of ties, motivation	Participative action research	Learning community	Wagenaar & Hulsebosch, 2008
	Seeding patterns	Descriptive case study	Strategic community	Kodama, 2001b

Table 9: Publications assigned to member inputs

2.3   Mediators

After putting input factors under the microscope for closer analysis, subsequently the main focus shifts to give closer attention to mediators which function as 'translators', in the sense that they transform inputs into outcomes and include emergent states and processes (Mathieu et al., 2008). Hitherto Tom is aware of the importance of organizational contexts on the functioning of organizational communities, as is Ina concerning community factors and Norm concerning member factors. However, they might be interested in how the 'inner' mechanisms of organizational communities transform these inputs into outcomes. In other words, they may be interested in how their inputs are transformed into performance of the community. The following section describes these mechanisms and may give Tom, Ina, and Norm the needed background information to understand community mechanisms in greater detail.

2.3.1   Emergent states

Tom, Ina, and Norm may be particularly interested in states that describe the emergent structure of communities or, in other words, psychological constellations among community members. These emergent states receive extensive attention in scholarly works. The following section displays and summarizes identified emergent states. They include proximity, strength of ties, centrality and stickiness. The subsequent table (next page) condensates key findings from literature. For an overview, Tom, Ina, and Norm find all studies assigned to these states in the table at the end of this section including corresponding studies, community focus, methods applied, related topics, and content focus.

According to Hussler & Rondé (2007, p. 300), **proximity** plays a crucial role in organizational communities. Proximity describes relationship of two or more community members in terms of their closeness or proximity. Two means of proximity are frequently discussed in community literature: 1) geographic and 2) cognitive proximity.

First, *geographic proximity* is defined as the spatial closeness of community members (Amin & Roberts, 2008, p. 366). Studies indicate that geographic proximity favors knowledge transfer, especially of tacit knowledge in situations when community members do not have a shared understanding (Hussler & Rondé, 2007, pp. 290, 300). For instance, Hussler & Rondé (2007, pp. 290, 300) are able to show that researchers belonging to one community should meet personally at the beginning of their collaboration to generate a shared understanding. By contrast, communities of practice have no particular need to meet in person as enough shared understanding is present by the practices they overlap. Second, *cognitive proximity*, often used synonymously to homogeneity, describes the mental closeness of community members. It includes sharing similar objectives and knowledge bases as well as mutual understanding (Nooteboom, 2000, p. 71; Boschma, 2005; Knoben & Oerlemans, 2006, p. 77). Cognitive proximity may result from joint practices as described above, similar professional backgrounds and mindsets or even hobbies. Nooteboom (2000, p. 71) emphasizes the janiform nature of cognitive proximity, as "[...] large cognitive distance has the merit of novelty but the problem of incomprehensibility" (Amin & Roberts, 2008, p. 365; Hussler & Rondé, 2007, p. 300; MØrk et al., 2008, p. 12).

In sum, research shows that a minimum of cognitive proximity among community members has to be achieved to enable communication between otherwise cognitively distant community members to ensure effective collaboration (Hussler & Rondé, 2007, p. 289). However, Amin & Roberts (2008, p. 365) demand a greater focus on cognitive proximity in scholarly works.

Proximity describes the *closeness* of two or more community members, e.g. in terms of spatial (geographic) closeness or mental closeness (homogeneity) [between person on the right and on the middle]. Community members that are not close but far away, either geographically or in mindset, may be described as distant, i.e. spatial distance or cognitive distance (heterogeneity) [as between the person in the middle and on the right]

 Cognitive proximity enables for efficient collaboration, whereas *cognitive distance* has advantages concerning novelty (innovation potential)

 Strength of ties characterizes the intensity of interactions (frequency, openness, and duration) among community members. Ties may be *strong* [symbolized by thick arrows between the persons left and upper right], *weak* [between person on the left and lower middle] or something in between [between person on the lower middle and upper right]

 Strong ties among community members positively influence cohesion, whereas *weakly connected* community members may bring in more and new ideas (innovation potential)

 Centrality refers to the role a community member plays in a specific community. A community member may be considered as *central* if he/she has internalized community values, norms, perspectives, etc., and executes important tasks [persons close to the red dot in the middle]. Community members that still try to understand the community and its values, norms, and perspectives may be described as *peripheral* [persons at the corners upper and lower left, and upper and lower right]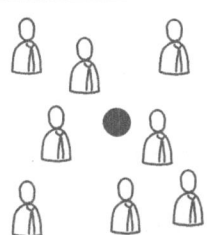

 Central community members are important for the stability of a community, whereas *peripheral* community members bring in new ideas, experiences, and perspectives (innovation potential)

 Stickiness describes the ease with which knowledge is exchanged from one community member to another. If knowledge is *sticky*, it is only exchanged with considerable effort or cost [knowledge is not transferred from the left to the right person's crossed-out arrow in the upper middle]

 Stickiness reduces exchange of knowledge and creates refusal of new ideas

Table 10: Emergent states: Proximity, strength of ties, centrality and stickiness

Next, well aware that relationships among community members vary in terms of their intensity, Tom, Ina, and Norm are confused about how they influence community activities and outcomes. This phenomenon has recently been studied under the term 'strength of ties'. Building on social network theory, strength of ties is characterized by the intensity of interaction, frequency of interaction, openness of communication, and duration of interaction (Granovetter, 1973; Bogenrieder & Nooteboom, 2004, p. 292). It is frequently separated into 1) strong ties and 2) weak ties (Granovetter, 1973).

First, *strong ties* are characterized by intensive, frequent, and long-lasting interactions applying an open communication style (Granovetter, 1973) among community members. In other words, those community members entertaining intensive collaboration for many months or years may be characterized as having strong ties whereas those community members who start interacting may be described as having weak ties. Bogenrieder & Nooteboom (2004, p. 294) find that strong ties foster cohesion, due to feelings of mutual bonds and feelings of trust, such as reciprocity beliefs. On the other side of the coin 'group thinking' and 'lock-in' may also result from strong ties (Bogenrieder & Nooteboom, 2004, p. 295). Second, *weak ties* constitute irregular, loose, and spontaneous interactions. They allow inclusion of ideas and knowledge stocks which are often completely new to the community, as they cross traditional boundaries and worldviews (Assimakopoulos & Yan, 2006, p. 16; Björk & Magnusson, 2009, p. 669).

Results from empirical studies imply that organizational communities should foster strong ties to realize cohesion and trust, and at the same time stay open for new members or peripheral contributors to benefit from ideas and knowledge from varying backgrounds. However, Björk & Magnusson (2009, p. 669) emphasize, that "[...] the ideation process taking place in groups stands out as more complex and appears to be harder to understand and explain with the external connections to the network than for individuals" and thus needs further research.

Moreover, community members vary in their position within the community. Whereas some community members play a central role for the community, others experience a more peripheral role. This phenomenon, which originates from Brown & Duguid's (1991, p. 48) conception of legitimate peripheral participation, is coined **centrality** and describes the position of community members (MØrk et al., 2008, p. 14; Davies, 2005, p. 565). Two major means of centrality are found: 1) absorption and 2) integration.

First, *absorption* refers to the degree to which community members internalize community perspectives, ideas, objectives, values, and norms (Witzeman et al., 2006, pp. 528-529). One major incentive to adopt these community features refers to gaining status. Along with higher status community members broaden their scope of opportunities to influence community activities. Research shows that community members who are less central, i.e. who have not yet absorbed community features, are only given minor tasks. Conversely, central community members influence the community, for instance because they are accepted project managers. In sum, community members adopting community features move closer to the core of the community or in other words become more central (Witzeman et al., 2006, pp. 529–530). Second, *integration* describes inclusion of perspectives, ideas, objectives, values, and norms that stem from outside the community or are to some extent contrary to community features. Research hints at the interdependent relationship of absorption and integration: While community members become more and more central through absorption, they adapt to community objectives, etc., and hence are less able to integrate new ideas, experiences, etc. On the other hand, their values, norms, etc., are increasingly embedded in the community's shared understanding (Hislop, 2003, p. 181).

Main topic:	Related topics:	Applied method:	Focus:	Study:
Proximity	Knowledge process, learning process, trust, outcome, stickiness	Theoretical	Community of practice	Kerr & Jermier, 1978
	Knowledge process	Quantitative case study	Knowledge community	Hussler & Rondé, 2007
	Knowledge process, structure, seeding patterns	Qualitative case study	Community of practice	Kasper, Mühlbacher, & Müller, 2008
	Knowledge process, weak ties, stickiness, seeding pattern	Quantitative questionnaire	Community of practice	Bertels, Kleinschmidt, & Koen, 2011
	Outcome, learning process	Participative action research	Community of practice	MØrk et al., 2008
Strength of ties	Boundary objects	Participative action research	Community of practice	Assimakopoulos & Yan, 2006
	Proximity, boundary objects, trust, outcome, social process, motivation	Qualitative case study	Learning community	Bogenrieder & Nooteboom, 2004
	Outcome, proximity	Social network analysis	Innovation community	Björk & Magnusson, 2009
	Status, social process, structure, strategy	Qualitative case study	Innovation community	Kirschten, 2006
	Centrality, boundary objects, structure	Theoretical	Innovation community	Lakhani & Hippel, 2003
	ICT, structure	Social network analysis	Community of practice	Tyler, Wilkinson, & Huberman, 2005
	Strength of ties, structure, motivation, strategy	Descriptive case study	Innovation Community	Fichter, 2006b
Centrality	Proximity, trust, social process	Qualitative case study	Learning community	Kinnear & Taylor, 1983
	Boundary objects, learning process	Theoretical	Learning community	Witzeman et al., 2006
	Outcome, learning process, boundary objects, proximity, knowledge process	Qualitative case study	Community of practice	Hislop, 2003
	Boundary objects, outcome, motivation, trust	Theoretical	n.a.	McMillan & Chavis, 1986
	Learning process, knowledge process, social process	Descriptive case study	Community of practice	Lave & Wenger, 1991
	Trust, boundary objects, proximity, outcome	Descriptive case study	Innovation community	Lamnek, 2000
	Proximity, outcome, social process	Qualitative case study	Innovation Community	van Oost et al., 2009
	Proximity, strength of ties, social process	Qualitative case study	Community of practice	Borzillo, Aznar, & Schmitt, 2011
	Status, strength of ties	Qualitative case study	Community of practice	Davies, 2005
Stickiness	Trust, knowledge process	Qualitative case study	Knowledge community	Siebert, 2006
	Motivation, trust, knowledge process	Quantitative questionnaire	Knowledge community	Ancona & Caldwell, 1992
	Trust, motivation, culture, structure, knowledge process, social process	Theoretical	Knowledge community	Desouza et al., 2008
	Social process, structure, trust, status	Qualitative questionnaire	Community of practice	Sawhney & Prandelli, 2000

Table 11: Publications assigned to emergent states

In this sense, centrality is understood as an emergent state in which community members continuously develop from peripheral to full membership. Hence, organizational communities constantly include new ideas, perspectives, and experiences, if influx of new members is maintained (Crossan, Lane, & White, 1999).

Lastly, Norm knows from his own experience that knowledge is sometimes hard to exchange. This **stickiness** of knowledge is defined as "[...] the incremental expenditure required to transfer that [a given] unit of information to a specified locus in a form usable by a given information seeker. When this cost is low, information stickiness is low; when it is high, stickiness is high" (Hippel, 1994, p. 430). In other words, the harder knowledge exchange between community members is, the stickier it is. Two major means of stickiness are identified: 1) individual resistance and 2) prohibitive situation.

First, *individual resistance* refers to community members who are not willing to exchange knowledge. Most importantly it is found that if motivation or trust is lacking this inhibits knowledge exchange (Szulanski & Cappetta, 2003, p. 523; Andrews & Delahaye, 2000, p. 800; Szulanski, 2000, p. 12). For instance, Szulanski & Cappetta (2003, p. 523) find that community members who do not see any personal benefit are unlikely to actively exchange knowledge. Second, *prohibitive situations* are social conditions in which community members do not feel at ease (Szulanski & Cappetta, 2003, p. 525). They are "[...] typified by hesitancy, stubbornness, awkwardness, and unpleasantness" (Szulanski & Cappetta, 2003, p. 514). Examples of such social conditions include a lack of self-efficacy among community members or a poor atmosphere within the community.

Stickiness is seen as a major reason for failed knowledge exchange and rejection of new ideas (Katz & Allen, 1982). However, until now it has remained underexplored how prohibitive situations could be prevented to overcome stickiness in organizational communities. *Processes*

As Tom, Ina, and Norm already have first impressions about the 'inner mechanisms' of organizational communities, i.e. emergent states, they want to know more about how knowledge is actually exchanged, social interactions take place, and learning actually unfolds. The following sections put together relevant insights from organizational community literature concerning knowledge, learning and social process. The following table recapitulates key aspects identified from literature on processes. Additionally, the table at the end of this section shows content focus, related topics, applied method, community focus and referred study.

A considerable number of publication in the area of organizational communities deals with **knowledge processes**. Researchers agree that supportive knowledge processes are a major source to generate outcomes such as innovations (Argote et al., 2000, p. 2; Nonaka & Takeuchi, 1995; Reinmoeller & Chong, 2002, p. 165). Knowledge processes may be defined as the articulation, reconciliation, and application of knowledge in social contexts in recurrent cycles (Nonaka & Toyama, 2003; Reinmoeller & Chong, 2002, p. 168). Accordingly, knowledge processes are best described in terms of 1) socialization, 2) articulation, 3) reconciliation, and 4) application (Peltonen & Lämsä, 2004, p. 252; Palincsar et al., 1998, p. 10; Sawhney & Prandelli, 2000, p. 46; Bechky, 2003, p. 322; Østerlund & Carlile, 2005, p. 98; Lee & Williams, 2007, p. 513).

	Knowledge processes describes how knowledge is exchanged between community members. It includes *socialization*, i.e. the internalization of emotional and thought patterns [left arrow between persons], *articulation* of thoughts [arrow from thought to speech bubble on the left], *reconciliation*, i.e. mutual updating [upper right] and *application* [lower right]	
	Successful *knowledge processes* are crucial to realize given potentials, such as in terms of innovation or task achievement	
	Learning processes may best be understood as processes through which community members develop minor or major parts of their *identity*, including professional or social aspects [persons from left person with a 'minus' to right person with a 'plus']. Acquisition of additional behavior *repertoire* through *consensus* among multiple community members [upper middle] is the modus operandi	
	Learning processes in communities unfold 'on the job' or in other words while solving a problem. They create rich understandings of complex interdependencies among multiple influencing factors	
	Social processes describe how cohesion emerges from democratic communicative processes among community members. It includes the creation of *awareness*, i.e. marketing of interests [left], *ignition* of activities, i.e. convincing other community members to contribute [left middle], *engagement* as actual behavior of several community members [right middle] and *confirmation*, i.e. validation of adequacy of solutions [right]	
	Social processes steer the direction of community members' activities. By this, social processes also bind community members together to fulfill one or a set of commonly accepted objectives	

Table 12: Processes: knowledge, learning and social processes

First, during *socialization*, knowledge is internalized in social contexts through action and perception (Østerlund & Carlile, 2005, p. 98; Nonaka & Toyama, 2003, p. 5). In other words community members gain knowledge mostly by updating their knowledge stocks for collaboration at a later stage. Second, *articulation* expresses accumulated knowledge in an understandable, comprehensive, detailed, and transferable manner, often in terms of stories, myths, practices or demonstrations (Østerlund & Carlile, 2005, p. 101; Brown & Duguid, 1991, p. 47). Brown and Duguid's reflection of Orr's (1986) example of technicians, trying to repair a copier machine at a customer's site, shows that all affected technicians shared their experiences by telling stories about earlier incidents and their interpretations. In this case example, telling these stories was a crucial element for the later successful repair of the machine. Third, *reconciliation* is a collaborative process, in which individual knowledge stocks are mutually transformed by previous articulations of other community members (Østerlund & Carlile, 2005, p. 101; Soekijad et al., 2004, p. 10; Palincsar et al., 1998, p. 10). It often includes iterative cycles of negotiation among community members. For instance, Brown & Duguid (1991, p. 44) show that technicians are able to solve severe problems with a machine when they reflect on and reconcile the articulations of other community members. Fourth, *application* of knowledge manifests if several community members apply the exchanged and developed knowledge in actual work. It is achieved if community members achieve consensus, internalize new insights and apply them (Lee & Williams, 2007, p. 513; Peltonen & Lämsä, 2004, p. 255; Nonaka & Toyama, 2003, p. 5). In the case of Orr's technicians, all three technicians tried the solution they created together and the machine they would have had to replace started to work again.

Communities receive considerable attention from scholars in the field of education but also in management as a tool for 'training on the job'. Scholarly publications on these issues are summarized under the term **learning processes**. Learning in communities is often labeled as 'collective learning in situ'. It has the benefit that learning is embedded in context and thus creates a better understanding of the complex interplay of multiple influences that enable positive outcomes (Brown & Duguid, 1991, pp. 47–50). The following three major means are identified as most important for learning processes in organizational communities: 1) acquisition of repertoire, 2) change of identity, and 3) consensus (Brown & Duguid, 1991, p. 48; Wilson, Goodman, & Cronin, 2007, p. 1044, Wilson et al., 2007, p. 1045).

First, community members *acquire additional repertoire* through social discourse of diverse interpretations, varying perspectives and worldviews among community members (Wang & Ramiller, 2009, p. 714; Novicevic et al., 2007, p. 374; Weick, 2002, p. 58). Additional repertoire mainly refers to an individual's gain of knowledge or of behavior repertoire (Wang & Ramiller, 2009, pp. 714–716). Second, community members *change their identities* while learning in communities as they revise perceptions of self (Handley et al., 2006, p. 644; Weick, 2002, p. 58). Perception of self may include self-awareness of capabilities and psychological states or previously unconscious paradigms or self-conceptions. Change of identity is likely to emerge when community members with different or sometimes conflicting perspectives and worldviews interact as they challenge the accepted (Novicevic et al., 2007, p. 374). Third, it is only if *consensus* is achieved – that is, a shared interpretation that integrates differences and creates rich understandings – that the learning processes may be seen as successful (Wang & Ramiller, 2009, p. 714; Novicevic et al., 2007, p. 374; Handley et al., 2006, p. 645; Wilson et al., 2007, p. 1044).

Previous research on organizational communities agree that **social processes** may be seen as processes binding community members together (Brown & Duguid, 1991, p. 43; Wenger, 1998, pp. 7–9; Wenger et al., 2002, p. 10; Drath & Palus, 1994; Palincsar et al., 1998, p. 17). Summarizing literature, social processes are described as (mainly) democratic communicative processes which manifest and stabilize communities and create cohesion (Swan et al., 2002, p. 492; Linehan & McCarthy, 2001, p. 146). Social processes include four major means: 1) awareness, 2) ignition, 3) engagement, and 4) confirmation (Fox, 2000, p. 862; Swan et al., 2002, p. 492; Linehan & McCarthy, 2001, p. 146; Fiol & O'Connor, 2002, p. 540; Kodama, 2001a, p. 1078).

First, *awareness* describes community members putting forth an issue, for instance issues with high importance, in a recognizable way while offering a set of actions to address this issue (Fox, 2000, p. 862; Swan et al., 2002, p. 492; Gongla & Rizzuto, 2001, p. 847; Wenger et al., 2002, p. 69). More descriptively, community members emphasize the importance of an important issue and try to convince others. In short they create awareness among community members. Second, community activities are *ignited* as soon as community members are willing to (temporarily) engage (Fiol & O'Connor, 2002, p. 539; Fox, 2000, p. 862). In short, community members accept the need to solve a specific issue and are willing to spend their time and energy or sometimes even money. Third, *engagement* includes actual collaborative activities from these members (Swan et al., 2002, p. 492). They act around issues and solve them by interacting intensively. In terms of Orr's (1986) example, technicians choose to solve the problem with the machine and start to update each other with their information and stories. They do not choose the second option: to simply replace the machine (going along with a loss in reputation). Fourth, *confirmation* validates the solution found for an issue and disseminates these results for wide application within the community (Swan et al., 2002, p. 492; Fiol & O'Connor, 2002, pp. 541–542). The technicians of the example above noted their solution in the maintenance manual to be available for other technicians with similar issues. Hence, social processes are an important means by which community activities are organized, in the sense that they steer activities among community members (Linehan & McCarthy, 2001, p. 146; Fiol & O'Connor, 2002).

Main topic:	Related topics:	Applied method:	Focus:	Study:
Knowledge process	Proximity, strength of ties, outcome, boundary objects	Theoretical	Community of practice	Amin & Roberts, 2008
	Proximity, seeding patterns	Qualitative case study	Community of practice	Soekijad et al., 2004
	Social process, learning	Conceptual	Community of practice	Peltonen & Lämsä, 2004
	Motivation	Quantitative questionnaire	Virtual community	Kim, Song, & Jones, 2011
	n.a.	Theoretical	Knowledge community	Lindkvist, 2005
	Seeding patterns, proximity, strong ties	Qualitative case study	Community of practice	Palincsar et al., 1998
	Seeding patterns, outcome, objectives, proximity	Theoretical	Innovation community	Sawhney & Prandelli, 2000
	Boundary objects, social process, stickiness, outcome	Theoretical	Community of practice	Brown & Duguid, 2001
	Boundary objects, proximity, learning process	Participative action research	Community of practice	Bechky, 2003
	Centrality, strength of ties	Theoretical	Knowledge community	Østerlund & Carlile, 2005
	Centrality, stickiness	Theoretical	Knowledge community	Lee & Williams, 2007
Learning process	Knowledge process, outcome, boundary objects, centrality	Theoretical	Learning community	Wilson et al., 2007
	Knowledge processes, ICT,	Theoretical	Community of practice	Brown & Duguid, 1991
	Proximity, social process, knowledge process	Theoretical	Community of practice	Novicevic et al., 2007
	Social process, proximity	Theoretical	Learning community	Weick, 2002
	Boundary objects, proximity, social process, knowledge process, motivation	Qualitative case study	Learning community	Wang & Ramiller, 2009
	Structure, boundary objects, centrality, outcome	Theoretical	Community of practice	Handley et al., 2006
	ICT, knowledge process, status	Conceptual	Learning community	Armstrong & Fukami, 2010
	Social process	Qualitative case study	Community of practice	Lave & Wenger, 1991

Social process	ICT, centrality, status, strength of ties	Participative action research	Community of practice	Barab et al., 2003
	Boundary object, strength of ties, proximity, structure	Ethnographic study	Community of practice	Kavanagh & Kelly, 2002
	Strategy, structure, status, boundary object	Qualitative case study	Strategic community	Kodama, 2001a
	Motivation, knowledge process, learning process, outcome	Qualitative case study	Community of practice	Linehan & McCarthy, 2001
	Boundary object, knowledge process, outcome	Theoretical	Community of practice	Mutch, 2003
	Knowledge process, boundary object	Ethnographic study	Community of practice	Patriotta, 2003
	Outcome, strength of ties, knowledge process, boundary objects, proximity, motivation	Qualitative case study	Community of practice	Swan et al., 2002
	Structure, motivation,	Qualitative case study	Community of practice	Gongla & Rizzuto, 2001
	Learning process, status, ICT	Theoretical	Community of practice	Fox, 2000
	Learning process, knowledge process, boundary objects, stickiness	Qualitative case study	Community of practice	Fiol & O'Connor, 2002
	Knowledge process, learning process, status, centrality, motivation	Conceptual	Community of practice	Drath & Palus, 1994

Table 13: Publications assigned to processes

2.4   Outcomes

Tom, Ina, and Norm now possess updated knowledge on organizational communities, specifically concerning organizational, community, and member input factors (factors that influence mediators and outcomes) and emergent states and processes as mediators (transforming inputs into outcomes). The only missing piece of the puzzle relates to created outcomes of organizational communities. Despite manifold hints concerning the influence of specific factors on the quality of outcomes, surprisingly few studies focus specifically on the outcomes of organizational communities. Still, they provide interesting insights concerning the effectiveness of these communities. The following table condensates and visualizes knowledge from literature. Moreover, Tom, Norm, and Ina find a summary of studies focusing on outcomes in the following section. Table 6 shows publications, community focus, methods applied, related topics, and content focus.

Qualitative outcomes [right] refer to decreased learning curves, i.e. community members learn faster, increased timeliness of problem responses, prevention of rework and increased innovativeness	
Qualitative outcome indicators suggest that communities may be a major source of increased productivity	
Quantitative outcomes relate to measurable indicators that show the effects of community-related activities. It includes direct savings as well as calculations concerning response times or the application of scholarly constructs, such as social capital	
Quantitative outcome indicators support the notion that community-related activities increase productivity in several measures, e.g. achieved savings in work hours or additional revenues, as well as the number of innovations created	

Table 14: Outcomes: Qualitative and quantitative approaches

Measuring **outcomes** is especially important in the context of organizational communities, as they lie outside the organization's hierarchical control and are thus not captured by traditional performance measures (Wenger & Snyder, 2000, p. 145). Outcomes are defined by measures trying to estimate the value of organizational community activities (Millen et al., 2002, pp. 72–73). They are measured using 1) qualitative or 2) quantitative approaches.

First, *qualitative approaches* show that organizational communities effectively decrease learning curves, facilitate fast responses, prevent rework, and generate continuous as well as discontinuous innovation (Wenger & Snyder, 2000, p. 139; Jakubik, 2008, pp. 20–21;

Millen et al., 2002, pp. 70–72). Numerous examples show how community approaches especially facilitate the development of innovations. Second, Wenger (2000, p. 145), for instance, reports that Shell has saved two to five million dollars and generated additional revenue of thirteen million dollars by community activities, applying a *quantitative approach*. Moreover, Shah et al. (2006) are able to demonstrate that innovation development applying community approaches is superior to established R&D approaches. Additionally, knowledge and social processes are evaluated using scientific constructs, e.g., social capital (Verburg & Andriessen, 2006, pp. 20–21; Dewhurst & Cegarra Navarro, 2004, p. 328).

Beyond articles in this category, around half of all articles in this literature review refer to innovation outcomes. Some examples include articles concerning proximity, strength of ties, and centrality (Nooteboom, 2000, p. 71; Amin & Roberts, 2008, p. 365; Bogenrieder & Nooteboom, 2004, p. 295; Crossan et al., 1999). All articles hint at organizational communities as a major source of innovations.

Main topic:	Related topics:	Applied method:	Focus:	Study:
Quantitative	Motivation, social process, seeding patterns, knowledge process	Descriptive case study	Community of interest	Armstrong & Hagel III, 1996
	Knowledge process, culture, decentralization	Quantitative questionnaire	Community of practice	Dewhurst & Cegarra Navarro, 2004
	Strength of ties, boundary objects, trust, proximity	Qualitative questionnaire	Community of practice	Lesser & Storck, 2001
	Knowledge process, objectives, boundary objects, ICT	Qualitative questionnaire	Community of practice	Verburg & Andriessen, 2006
Qualitative	Trust, status, knowledge process, social process	Qualitative questionnaire	Community of practice	Millen et al., 2002
	Seeding patterns, purpose	Descriptive case study	Community of practice	Wenger & Snyder, 2000
	Knowledge process, learning process, proximity	Participative action research	Community of practice	Jakubik, 2008

Table 15: Publications assigned to outcome

2.5 Discussion: Organizational communities

Thus far, this literature review has updated Tom's, Ina's, and Norm's knowledge on organizational communities. In other words, it has provided a contemporary analysis of scholarly publications and comprehensively summarized key definitions, factors, and aspects. Based on 131 identified contributions, this study has examined and distilled thematic focal points, as well as theoretical and empirical results. Moreover, the analysis has revealed supplementary insights concerning gaps in research and avenues for future research. To advance clarity and coherence in organizational community research, the subsequent section organizes gaps in and future avenues for research following the input, mediator, and outcome framework.

2.5.1 Input factors

As Tom has gained considerable information about **organizational contexts** and their influence on organizational communities, he is especially interested in those aspects with which researchers are probably not able to help him. Two major gaps in current research are identified within the category of organizational context: 1) the interdependencies of organizational contexts and 2) the strategies to create supportive contexts for organizational communities. First, considerable knowledge is provided by contemporary research concerning the antecedents of organizational contexts. For instance, Kodama (2005a) shows how organizational communities may serve the strategic objectives of an organization. Other authors also demonstrate that, despite their democratic and self-governing structures, organizational communities are still dependent on structural support, such as time and available budget, as well as sponsors to support activities (Brailsford, 2001; Tarmizi & de Vreede, 2005). Also, the organizational culture, in terms of awareness and recognition, is shown to be a requirement for organizational communities to flourish (Brazelton & Gorry, 2003, p. 25; 2006, pp. 127–128). Lastly, the provision of ICT infrastructure is also seen as crucial.

These empirical and theoretical findings are meaningful not least because they counter-intuitively show that communities in organizations are by no means 'self runners' but have to be supported by organizational contexts. Even though academics identify several candidates that may provide supportive organizational contexts, it remains rather unclear how they influence organizational communities and how they relate to mediators and outcomes. For instance, Kimble & Bourdon (2008, p. 466) state that "[...] getting people into sharing mode is the hardest part." Besides showing that organizational culture has a positive influence on organizational communities generally, insights concerning specific antecedents to actually get people involved through cultural factors are not provided. Hence, it remains rather underexplored how organizational context influences mediators and outcomes, specifically concerning their influence on innovation-related outcomes. Moreover, most studies use 'snapshots', i.e. static approaches mostly collecting data at one point in time, to analyze contexts for organizational communities. However, organizational contexts that support organizational communities do mostly not emerge randomly, but are a result of systematic adaptation. Until now, the analysis of strategies to consciously adapt organizational contexts to facilitate organizational communities has not been a major focus. To the best of the author's knowledge, no research is available to date that derives strategies to consciously adapt organizational contexts for the sake of organizational communities.

Hence, at present, little is known regarding how specific organizational contexts influence outcomes and how organizational contexts may be adapted. This gap in research offers fruitful opportunities to advance knowledge about organizational (innovation) communities considerably and to include organizational communities in future research agendas. In sum, literature on organizational contexts presents meaningful and interesting insights on the interrelation of organizational context and organizational communities. Building on 'these strong shoulders', studying the role of organizational contexts on mediators and outcomes and strategies to alter these contexts may add interesting insights. Organizational behavior theories may provide inspirations to resolve these research questions. For instance, structuration theories (Jarzabkowski, 2008) may offer a sound theoretical foundation to answer questions on integrating organizational communities into organizations' contexts.

A considerable number of scholars study the important role of **community contexts** on organizational communities. For instance, scholars study the fascinating role of boundary objects in organizational communities with great ambition and effort (Plaskoff, 2003; Bechky, 2003; Garrety et al., 2004). For instance, Bechky (2003) give rich insights in how boundary objects function to create shared meanings among employees with different professional backgrounds at a manufacturing site. However, opportunities for further research exist especially concerning how these boundary objects develop dynamically over a longer period of time, what kinds of boundary objects are specifically applicable and how boundary objects influence innovation development, especially in online settings.

Moreover, studies show that community contexts have to be carefully balanced, as they may fuel momentum but on the other side of the coin may hamper community activities (Breu & Hemingway, 2002; Probst & Borzillo, 2008). For instance, providing a facilitator helps to keep the community thriving (Pemberton et al., 2007, pp. 67–68). However, facilitators may also 'do too much of a good thing', undermining the democratic structure of organizational communities and thus constraining outcomes. Consequently, longitudinal studies may be specifically applicant to study the role of boundary objects and their influences on mediators and outcomes. To tackle Ina's main issue a sensemaking perspective (Weick, 1988; Weick, 1995) could be particularly helpful for investigating these interrelations particularly if innovation is at focus (Christiansen & Varnes, 2009).

Member factors have also received considerable attention in scholarly works, especially topics such as motivation, trust, and status. For instance, in-depth knowledge is created concerning the motivation structure of community members (Jian & Jeffres, 2006, pp. 242–243; von Krogh et al., 2003, p. 1235; Komito, 1998). This is similarly true for trust (Soto et al., 2007, pp. 354–355; Hsu et al., 2007, pp. 157, 160) and status aspects (Fleming & Waguespack, 2007, p. 165; Lerner & Tirole, 2002; Hippel & von Krogh, 2003).

However, application of a *wider spectrum of theoretic perspective* may offer opportunities to extend knowledge in organizational community literature. For instance, applying theories from psychology that go beyond motivation structures may be able to explain individual behaviors more accurately. Some candidates include flow theories but also theories that explain the psychological functioning of individuals, such as social cognitive theories or emotion theories.

Moreover, besides selective insights little knowledge is available concerning *interdependencies* between individual factors and mediators or outcomes. Specifically, interdependencies among factors such as motivation, trust, and status, but also between member aspects, mediators, and outcomes are rarely explored. Hence, only vague and incomprehensive understandings of their influences on mediators or outcomes are available. For instance, it is widely accepted that motivation, trust, and status do not exert influence on mediators and outcomes independently (Hsu et al., 2007, pp. 165–166; Franke & Shah, 2003, pp. 172–173). However, little is known on interdependencies of varying levels of motivation, trust and status and mediators and outcomes (Muller, 2006, p. 396). Thus, further studies could investigate the influence of varying levels of interdependent individual factors on mediators or outcomes.

2.5.2 Mediators

The analysis of **emergent states** brings about fascinating and meaningful insights. For instance, studies show that weak ties among community members are specifically useful if creative solutions are needed. On the other hand, strong ties seem to facilitate efficient task fulfillment. In a similar vein, cognitive proximity helps to 'get things done', whereas cognitive distance has the merit of creative solutions (Björk & Magnusson, 2009, p. 669; Bogenrieder & Nooteboom, 2004, p. 295; Nooteboom, 2000, p. 71). These findings help to understand why community members may be able to solve issues quickly at one point in time or may be able to develop innovative solutions at another point in time.

However, despite these advancements in understanding community mechanisms, some *contradictions* remain unsolved. For instance, cognitive proximity and strong ties support efficient task fulfillment but hamper creativity. Vice versa, cognitive distance and weak ties support creativity but hamper efficient task fulfillment. Exploring how creativity and efficiency may be achieved at the same time, i.e. organizing weakly connected or cognitively distant community members efficiently, may be especially interesting. To do so, a sensemaking perspective may provide the theoretical toolkit to study these contradictions, as it enables an examination of social interactions under the microscope. Yet again, longitudinal studies of these complex interactions may be a promising way to find mechanisms that facilitate efficiency and creativity at the same time.

2.5.3   Outcomes

One major weakness of organizational community literature may be seen in the lack of key performance indicators or key dependent variables to measure **outcomes** consistently. Today studies are rarely comparable as outcomes are not consequently reported. Consequently, cumulative knowledge building is hampered. Specifically, the majority of studies in organizational community literature remain generic when defining dependent variables, e.g., the 'viability of community'. Accordingly, scholars are encouraged to specify achieved outcomes explicitly. For instance, one of the outcome variables, long propagated by scholars but which has not been followed up, concerns organizational communities and their influence on *innovation outcomes*. Along other topics, Brown & Duguid (1991; 2001) emphasize that organizational communities are an important means to facilitate innovation outcomes. In addition, more than half of all articles in this literature review relate their findings to innovation outcomes. Hence, one major gap in research concerns clear definition of dependent variables, such as innovation outcomes.

2.5.4   Summary

The following table (next page) summarizes future avenues for research based on the discussed gaps within each factor. In other words, here Tom, Ina, and Norm can find those topics for which researchers have not yet found sound answers. The table provides the following information: 1) factor, 2) sub-topic, 3) research agenda, and 4) meaningful future avenues. First, *factor* describes one factor of the input, mediator, and outcome in one row. Following the structure of the literature review, the first three rows display the input factors organizational context, community context, and member; the following two rows show the mediator factors, processes, and emergent states; and the final row outcomes. The second array displays the *sub-topics* identified and presented in the literature review. For instance, on the organizational level strategy, structure, culture, and ICT are identified as major sub-topics, as has already been described. Third, the *main foci* of contemporary literature are presented, i.e. the research agenda, in the third array. Fourth, *meaningful future avenues* for research are presented in the last array. In sum, the following table summarizes the presented sub-topics, factors, dominant research agenda, and a selection of areas which deserve more research.

This part aims so far at identifying relevant studies in organizational community literature, summarizing and organizing literature, using an input, mediator and outcome framework and to present research gaps. In doing so, this review asks for more deliberate use of theoretical foundations and a better integration of studies and presents avenues for future research. Based on first results from literature, the following discussion of contents in the field of boundary-less communities adds to the presented knowledge. This is necessary as research in the field of organizational communities provides considerable knowledge concerning the application of ICT. Hence, to not run the risk of 'just' replicating existing knowledge from boundary-less communities, this stream of literature has to be considered. A more detailed presentation of contents of this stream of literature may be found in annex B. In sum, the identification of crucial gaps in research concerning organizational innovation communities builds on organizational community literature as well as on boundary-less community literature to address this issue.

Factor	Sub-topics	Research agenda	Meaningful future avenues
Organizational context	Strategy, structure, culture, ICT	Considerable advances in generically understanding influences of strategy, structure, culture and ICT. Exclusively qualitative and conceptual/theoretical research may be identified	First, research would benefit from a more nuanced understanding on the impact of specific organizational contexts on given mediators and outcomes. Second, analysis of strategies to alter organizational contexts in favor of organizational communities seems valuable
Community context	Boundary object, seeding pattern	Substantial research in this area is available, including how facilitators may fuel momentum, but also hinder community activities is available	Future research should consider the dynamic development of boundary objects and seeding patterns to derive recommendations concerning appropriate support
Member	Motivation, trust, status,	Increasing number of publication with a recent move form qualitative and theoretical towards quantitative approaches is detected. Substantial evolution of understanding motivational and trust-related aspects is identified	More research is admired, concerning interdependencies of member factors and interdependencies of individual variables and outcomes
Processes	Knowledge, learning, social processes	Considerable research in this area, specifically relating to knowledge processes, but increasingly studying social processes with a strong tendency towards theoretical and qualitative approaches is done. Increased popularity of studies concerning social processes in the recent years is detected	Research should consider the emergence and support of social processes in greater detail
Emergent state	Proximity, strength of ties, centrality, stickiness	Considerable upswing of research, especially concerning proximity and strength of ties is seen	Several contradictory findings were identified in current research. Future research could profit from resolving these contradictions, e.g. concerning the struggle between efficiency and effectiveness
Outcomes	Quantitative, qualitative	Relatively little research conducted, specifically identifying outcome variables and developing outcome measures is available	Future research would profit from explicitly defining outcome variables at large and more specifically including innovation outcomes as a dependent variable

Table 16: Summary of organizational community literature

2.6   Discussion: Boundary-less communities

Complementary to the literature review on organizational communities, boundary-less communities are also analyzed. The analysis is conducted because boundary-less communities – e.g. online communities, open source communities, and virtual communities – offer considerable insights concerning the functioning of communities in mainly online environments. For instance, researchers in this stream of community research provide rich understandings concerning the design of IT artifacts in support of communities: A considerable scope of scholarly work addresses the question of how motivational aspects may be fostered by providing IT features, such as star rankings, etc. Thus, this stream of research provides meaningful additional information to address the question of how to turn organizational innovation communities, which are considerable supported by ICT, into a management innovation. However, as the main focus remains on organizational communities, a detailed display of contents is not provided at this point. Still, articles being a matter of analysis are provided in Annex B and key findings as well as gaps and future avenues for research are discussed below, following the input, mediator, and outcome framework.[11]

2.6.1   Input factors

The **organizational context** includes ICT as the major driver. Building on a long tradition in studying ICT as a means to facilitate communities, an overwhelming scope of findings is identified. Major points of focus refer to (1) aesthetic design (Stanoevska-Slabeva, 2002; Chang et al., 2003; Boczkowski, 1999), (2) convenient navigation (Talukder & Yeow, 2007; Stanoevska-Slabeva, 2002), (3) facilitating social interactions (Talukder & Yeow, 2007; Wamalwa, 2007) and (iv) constantly adapting ICT to new developments (Chang et al., 2003; Stanoevska-Slabeva, 2002).

Besides understandings about designing ICT to facilitate information access, at present a minor weakness of literature refers to *the influence of specific functionalities* of ICT on other variables, such as participation and satisfaction in the category emergent state. Specifically, as ICT often differs in quality and functionality, findings may not be generalizable to a broader set of communities but may be bound to the specific context of studies. Hence, differences in ICT may account for bigger proportions in variance of findings compared to other variables as they offer alternative explanations for extracted causal relationships. Comparisons of findings across and testing acceptance and usability of different configurations of functionalities could advance research, as they dissolve existing uncertainty concerning the generalizability of findings. One strategy to resolve this issue may be to give detailed information about IT artifacts. This may include providing screenshots of artifacts to give readers rich understandings about the specific context.

Opportunities exist for including a broader range of **community contexts** to extend current knowledge. In general, empirical findings indicate that community contexts moderate various relations. For instance, boundary objects facilitate knowledge processes, social processes and participation mainly due to the shared understanding that is created among community members (Chang et al., 2007, p. 238; de Cindio et al., 2003, pp. 398, 400, 402; Wachter et al., 2000, p. 481). Moreover, seeding patterns support for instance social processes as facilitators organize relationships among community members. However, to date community factors have not been included as moderators.

Moreover, greater inclusion and exploration of community factors offer meaningful avenues to extent knowledge on communities in online contexts, specifically to explain contradictory empirical findings. Consequently, further empirical and theoretical research investigating latent constructs and their moderating and mediating variables as well as intermediating effects should be conducted (Leimeister et al., 2008, p. 386; Brown et al., 2007, p. 15). Boundary-less community literature could thereby profit from exploring issues inductively, such as applying grounded theory building. Inductive studies may have the merit of explaining more variance as of today. Given that boundary-less communities remain an important means to organize work, exploration and theory development are of interest both for researchers and practitioners. At best, deeper grounding might strengthen knowledge about boundary-less communities, display research gaps more explicitly and fuel broader theory development, also in related fields.

One major challenge concerning **member** factors stems from contradictory empirical findings. Even though scholars employ mature concepts stemming from motivation and trust research, some unexpected inconsistencies emerge from empirical investigations. For instance, empirical studies highlight motivational factors, such as improvements, reputation,

---

[11] Articles included in the literature review are made publicly available in the group 'online innovation communities' in Mendeley (www.mendeley.com).

and reciprocity as major drivers for participation (Ebner et al., 2009, p. 353; Lerner & Tirole, 2002, pp. 213–214; Quigley et al., 2007, p. 71; Harhoff et al., 2003). In contrast, Franke and Shah (2003, p. 173) find little support for this finding and suggest enjoyment and fun as major drivers. One may speculate that reasons for such contradictory findings result from low levels of explained variance or from findings bound to specific settings. However, one might also argue that the functioning of individuals in communities is more complex than in other contexts and therefore needs more elaborate concepts than those frequently applied. Following this interpretation, understanding member functioning in online communities may fuel theory building in psychology and other fields, and hence may be especially fruitful to explore. Hence, it might be beneficial to isolate a broader set of constructs, for instance from psychology, find specificities of these constructs in community contexts and feed the results back to related fields. In sum, online community contexts may be a beneficial context to refine constructs from related research fields.

2.6.2   Mediators

Boundary-less community literature concerns how **processes** unfold dynamically. Many scholars emphasize that, for example, knowledge and social processes are central to facilitate efficient task fulfillment and to produce valuable outcomes. Multiple concepts and theories are applied with the goal to shed light on the nature of these processes. Nonetheless, for the most part they do not sufficiently capture the dynamics of these processes, such as the discursive nature of knowledge and social processes. For instance, it remains unclear how innovations are actually developed within online environments in a collaborative way. One reason for this issue may lie in mostly asynchronous collaboration support. Boundary-less communities mainly rely on commenting or submitting preliminary solution for further refinement. For instance, in open source communities software code is submitted by one community member and refined by another. Yet, they do not actually collaborate in the sense that they work on the software code at the same time and discuss their insights. Rather, they download a piece of code, refine it and resubmit it for others to do the same. Hence, it remains a matter of speculation how the discursive nature of knowledge and social processes unfolds in boundary-less community contexts.

Moreover, whereas effects of strong and weak ties for efficient task fulfillment are well explored, little is known about the forming process of social ties in boundary-less communities and how this process may be supported (Brown et al., 2007, p. 15). Consequently, room for further exploration and theory building exists to explain if, and if so how, differences in interactions exert significant influence on outcomes. Particularly, exploring synchronous collaboration in such communities may add meaningful insights concerning how knowledge and social processes unfold and how they influence outcomes such as innovations. One possible perspective to study these dynamic processes may be seen in sensemaking theory. Sensemaking theory has already proved to be a useful perspective to explore dynamics at the intersection of social processes, knowledge processes, and (innovation) outcomes.

2.6.3   Summary

Before turning to a summary of avenues for future research, one important weakness of current boundary-less community literature has to be mentioned. Even though **methodological criticism** is not at the core of the discussion of literature in boundary-less communities, one methodological issue is worth noting. One important limitation of boundary-less community research to date concerns *selection bias* and particularly low response rates of online surveys. Response rates often remain at a low level: they seldom exceed 20 % response rates (Hars & Ou, 2002; Füller, 2006; Ke & Zhang, 2009), and can be as low as 3 % (Nonnecke, Andrews, & Preece, 2006). Low response rates are especially critical in boundary-less community literature, as community members differ in main characteristics considerably. For instance, Nonnecke (2006, p. 9) shows that more than 80 % of all respondents are those members who actively participate in the community, even though making up only a relatively small percentage of all participants (numbers typically vary from 5 % to 20 %; e.g., Beenen et al., 2004, p. 212). Hence, the internal validity of many findings might be questioned, as causal relationships may be distorted towards active participants and may not hold the test of selection bias. In other words, empirical findings may explain the variables of highly active community members adequately, but may not be applicable for the majority of community members. To control for selection bias, studies should account for the distorted distribution of respondents. Moreover, other approaches to collect quantitative data, such as telephone interviews that may result in higher response rates, should be considered.

The following table displays future avenues for research based on the above discussed gaps. The table thereby provides information concerning: 1) factor, 2) sub-topics, 3) research agenda, and 4) meaningful future avenues. The table therefore follows the same logic as the summary of organizational community literature. In sum, the following table summarizes the presented categorization, factors, dominant research agendas, and a selection of areas which deserve more research.

Factor	Sub-topics	Research agenda	Meaningful future avenues
Organizational context	ICT	Considerable advances in overarching influences and main functions of ICT in boundary-crossing communities are achieved. Studies apply a wide range of theoretical, qualitative, quantitative, and mixed methods	Research would profit from an ICT framework that captures dynamic developments of web-applications
Community context	Boundary object, seeding pattern	Substantial research in this area, including how facilitators may fuel momentum, but also hinder community activities is conducted	Future research should consider the dynamic development of boundary objects and seeding structure to derive recommendations concerning appropriate support
Member	Motivation, trust	Mainly quantitative approaches are applied. Substantial evolution of studies on motivational and trust-related aspects is confirmed	More research is admired to resolve contradictory empirical findings and to explain mediators and outcomes such as participation
Processes	Knowledge processes, social processes	Considerable upswing in research in this area, especially concerning social processes is detected. Strong tendency towards descriptive qualitative, but also conceptual and theoretical approaches is identified	Considerable scope for further studies exists, especially concerning truly collaborative activities (same time / same place) in online or ICT-enabled community
Emergent state	Strength of ties, participation, satisfaction	Substantial advances in explaining participation resulting from various variables is achieved. Research is dominated by quantitative approaches	Future research would profit from a statistical exploration of relationships of emergent states on processes and outcomes
Outcomes	Quantitative, qualitative	Considerable advances in measuring outcome variables, including self-ratings and ICT-based indicators, such as number of postings or concepts submitted is confirmed	Future research would benefit from developing a broader set of ICT-based indicators that capture actual behavior

Table 17: Summary of boundary-crossing community literature

3 Implications for organizational innovation communities

In the preceding, it has become clear that community research, whether in organizational or boundary-less contexts, offers an impressive number of meaningful contributions of manifold community aspects while applying diverse theories and constructs from various fields of research. An overwhelming scope of empirical as well as conceptual and theoretical findings exists in community research. Consequently, scholars can build on a long tradition of community research and a massive stock of knowledge concerning the functioning of communities not least as many studies apply narrative strategies to present findings.

However, the amazing scope and scale of literature makes a literature review especially challenging, as the rich findings have to be simplified and clustered into a few distinct categories that have to be described densely but simultaneously overarching. Facing these challenges, the comprehensive review of community literature (i) systematically organizes and summarizes the wide range of conducted studies, (ii) displays interdependencies among these studies, and (iii) derives with multiple avenues for future research. Consequently, the literature review may be seen as a kind of 'Yellow Pages' for community research. Researchers as well as practitioners such as Tom, Ina, and Norm will profit from these 'Yellow Pages' as they are able to search for relevant community aspects and to find further literature for more details. Moreover, has also shown how different studies connect with each other or, in other words, which studies are 'neighbors'. Lastly, based on the analysis of contents of publications and their interrelations future avenues for research are identified.

Based on the results of the systematic literature review, critical aspects for the transfer of innovation communities to the organizational context are identified. Hitherto, the analysis of community literature derives manifold avenues for future research. It would be impudent to claim that all gaps may be meaningfully addressed in one thesis. However, most critical gaps of research concerning organizational innovation communities may be identified and addressed. In so doing, this thesis contributes to transfer the phenomenon of innovation communities to the organizational context and therefore contributes to the establishment of a management innovation. In sum, most critical gaps in current research are presented based on the reflection of identified avenues for future research in community literature.

The identification of crucial gaps in research for organizational innovation communities is presented addressing Tom's, Ina's, and Norm's struggles. First, **Tom's struggle** concerning the appropriate design of organizational contexts and strategies to create these contexts is discussed. *This is what he learns from literature:* Organizational contexts foremost refer to structures and culture to support organizational communities. Briefly, he learns that organizational communities are dependent on structural support, such as time and available budget as well as sponsors to support activities (Brailsford, 2001; Tarmizi & de Vreede, 2005), and cultural support, in terms of awareness and recognition (Brazelton & Gorry, 2003, p. 25; 2006, pp. 127–128).

This is what he does not learn: Literature does not deliver clear assumptions concerning the role of structures and culture on innovation activities and outcomes. Moreover, he does not find hints in the literature on strategies to create organizational contexts in which organizational innovation communities are structurally and culturally anchored. *Hence, one major gap in research refers to the role of structural and cultural antecedents on innovation activities and outcomes and strategies to create such organizational contexts.*

Literature suggests that organizational innovation communities are dependent on structural and cultural support

Gaps in research concern influences of structure and culture on innovation activities and outcomes. Moreover, strategies to create supportive structures and cultures are missing

Table 18: Crucial gap in research concerning organizational integration

Second, **Ina's struggle** refers to understanding innovation development in organizational innovation communities and ways to facilitate this process. *This is what she learns from literature:* In short, literature on community contexts shows her that a facilitator may be effective to fuel momentum (Breu & Hemingway, 2002; Probst & Borzillo, 2008). Additionally, boundary objects are an important means to create shared understanding among community members (Plaskoff, 2003; Bechky, 2003; Garrety et al., 2004). From literature concerning mediators, she figures that weak ties and cognitive distance foster creativity in organizational communities but often hamper efficient task execution (Björk & Magnusson, 2009, p. 669; Bogenrieder & Nooteboom, 2004, p. 295; Nooteboom, 2000, p. 71) and that knowledge and social processes are crucial processes in communities (Brown et al., 2007, p. 15).

This is what she does not learn: Ina does not get detailed information about the nature of the facilitator's activities or the nature of boundary objects to positively influence innovation development in organizational innovation communities. She is also worried about including weakly connected and cognitively distant community members for the sake of creativity while accepting the handicap of potentially inefficient task execution. Moreover, she has hoped to gain knowledge concerning knowledge and social processes and how they unfold in online contexts in the pursuit of innovation. However, studying literature seems not to be of great help, potentiating her worries. *Hence, a major gap in research refers to the way social and knowledge processes unfold for the sake of innovation, especially if cognitively distant and weakly connected community members are integrated in these processes.*

Literature suggests that weakly connected and cognitively distant community members collaborating have significant innovation potential

Gaps in research concern understandings of how social processes for innovation development in organizational innovation communities unfold

Table 19: Crucial gap in research concerning social processes

Third, **Norm's struggle,** related to the engagement of employees in organizational innovation communities, is tackled. *This is what he learns from literature:* Reading the literature review, he sees that motivation (Jian & Jeffres, 2006, pp. 242–243; von Krogh et al., 2003, p. 1235; Komito, 1998), trust (Soto et al., 2007, pp. 354–355; Hsu et al., 2007, pp. 157, 160), and status influence the willingness of community members to contribute (Fleming & Waguespack, 2007, p. 165; Lerner & Tirole, 2002; Hippel & von Krogh, 2003). Literature, especially in the area of boundary-less communities, provides insights about how these factors may be implemented in an IT artifact.

This is what he does not learn: He stumbles across the fact that findings are sometimes inconsistent, leaving him puzzled with his struggle (Ebner et al., 2009, p. 353; Lerner & Tirole, 2002, pp. 213–214; Franke & Shah, 2003, p. 173). Additionally, he does not find an answer concerning his question of how to overcome the lack of confidence and fun constraining his innovation activities. Therefore he agrees with the call to identify and apply more elaborate concepts and theories to understand the psychological functioning of employees in organizational innovation communities. *Hence, a major gap in research refers to the understanding of individual engagement in organizational innovation communities.*

	**Literature** shows that motivation, trust, and status are crucial drivers for engagement. It is also shown how these factors may be implemented in an IT artifact.
	Gaps in research refer to the use of more elaborate concepts and theories to understand psychological functioning of employees in organizational innovation communities

Table 20: Crucial gap in research concerning engagement

Consequently, this thesis addresses these three most pressing gaps in research that result at the intersection of organizational community and boundary-less community literature. By juxtaposing Tom's, Ina's, and Norm's struggles and gaps in research, three major avenues for research are followed in the subsequent empirical parts with the aim to contribute to establishing organizational innovation communities as a management innovation:

- Analyzing the role of structural and cultural antecedents on innovation activities and outcomes and strategies to create such organizational contexts in support for organizational innovation communities (Part III).

- Exploring mechanisms of social processes in organizational innovation communities for the sake of innovation and ways to facilitate these processes (Part IV).

- Explaining individual's engagement in organizational innovation communities in the pursuit of innovation (Part V).

Consequently, the subsequent part sheds light on the first research avenue, answering the following main research question: "how does organizational integration influence innovation activities and outcomes of organizational innovation communities and how may organizational integration be altered?" This part derives with a taxonomy of organizational innovation communities, including four types of organizational innovation communities, and transition strategies to alter organizational integration. Tom learns how structural and cultural antecedents influence innovation activities and outcomes in organizational innovation communities and how these contexts may be established.

Part III A taxonomy of organizational innovation communities (empirical study I)

1 Setting the stage[12]

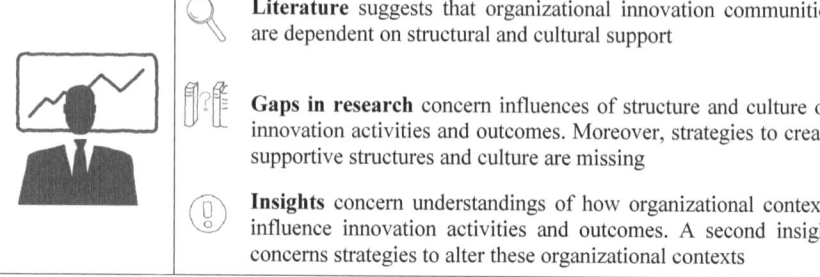

	Literature suggests that organizational innovation communities are dependent on structural and cultural support
	Gaps in research concern influences of structure and culture on innovation activities and outcomes. Moreover, strategies to create supportive structures and culture are missing
	Insights concern understandings of how organizational contexts influence innovation activities and outcomes. A second insight concerns strategies to alter these organizational contexts

Table 21: Taxonomy: Existing literature, crucial gap and insights

This part tackles the first major gap in scholarly research, i.e. missing understandings of the role of structural and cultural antecedents on innovation activities and outcomes and strategies to create such organizational contexts. In other words, Tom's major struggle to transfer the fashion of innovation communities to foster innovation development within organizations is addressed. He trusts the assumption, also supported by research, that organizational integration of such communities is a pre-condition for innovation to occur (Kodama, 2001b). As the gaps in contemporary literature on communities show, little is known about how organizational innovation communities need to be anchored to contribute significantly to innovation activities and outcomes within organizations. Previous research hints to a variety of reasons for failure of organizational innovation communities, ranging from lack of motivation (Fang & Neufeld, 2009; Roberts, 2006; von Krogh et al., 2003) to failed knowledge exchange (Bechky, 2003; Peltonen & Lämsä, 2004). To contribute to this discussion, this part focuses on the understudied link between organizational integration and its influence on innovation activities (i.e. the way individuals collaborate) and outcomes (i.e. the amount of innovations generated) of organizational innovation communities. Additionally, it identifies distinct types of transition strategies to anchor the organizational integration of organizational innovation communities. In so doing, Tom gains considerable information about antecedents of organizational contexts on innovation activities and outcomes and strategies to create these contexts.

To help facilitate this endeavor structuration theory is applied (Jarzabkowski, 2008, p. 622; Giddens, 1979, Giddens, 1984; Jones & Karsten, 2008). This theoretical perspective offers an instrument to determine the nature of organizational contexts and to discover transition strategies to adapt these contexts to given conditions for two reasons: First, it builds on the premise that organizational contexts frame individuals' activities and outcomes (Gladstein, 1984; Burns & Stalker, 1994; Lawrence & Lorsch, 1967). Specifically, several scholars emphasize the crucial role of supporting cultural and structural contexts for innovation activities and outcomes to occur (de Brentani & Kleinschmidt, 2004, pp. 312–314; Lawrence & Lorsch, 1967; Burns & Stalker, 1994). Second, organizational contexts are also shaped by individuals' activities characterized by choosing among multiple activity

---

[12] A previous version of part III is accepted for publication in the *International Journal of Knowledge-based Organizations* (Bansemir, Neyer, & Möslein, 2012). The article is co-authored by Anne-Katrin Neyer and Kathrin M. Möslein. Additionally, a considerably adapted version of part III is in press in the conference proceedings of the Informatik 2011 – Informatik schafft Communities (Bansemir, Neyer, & Möslein, 2011b).

opportunities (Giddens, 1979, Giddens, 1984; Orlikowski, 1996; Jarzabkowski, 2008, p. 622). In this regard, Jarzabkowski (2008, p. 622) hints at the effectiveness of interactive transition strategies (i.e. convincing others to engage) and procedural transition strategies (i.e. changing structural elements) to change given organizational contexts. Taken together, structuration theory is applied as the theoretical foundation to analyze both the influence of organizational contexts on organizational innovation communities and transition strategies to anchor organizational innovation communities within organizations.

This part is structured as follows: First the indicated theoretical perspective is presented, followed by an explication of the multiple case study methods applied. Then, four types of organizational innovation communities are identified, characterized by distinct innovation activities and outcomes. Next, the three identified transition strategies are presented. Finally, a taxonomy of organizational innovation communities, including 1) four types of organizational innovation communities and 2) transitions strategies to anchor such organizational communities within the organization, is derived.

2 Theoretical perspectives

 Structuration theory describes the following reciprocal relation: the organizational context, e.g. incentive systems, influence employees' behaviors AND employees' behaviors influence the organizational context, e.g. change of incentive system due to misbehavior.

 A gap in research concerns influences of structure and culture on innovation activities and outcomes

 Key insights: Structure [left], e.g. resources, and culture [middle, left], e.g. curiosity, play a crucial role for organizational integration of communities [middle, right] and innovation [right]

 Research question: How does organizational integration, by means of structural and cultural integration, influence innovation activities and outcomes of organizational innovation communities?

 A gap in research relates to strategies to create supportive structures and cultures

 Key insights: To increase organizational integration [move from left to right], two transition strategies are frequently highlighted: First, changing culture through interactions [middle, up] and second, by changing structures, e.g. incentive systems [middle, low]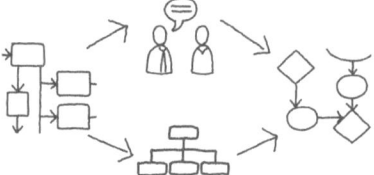

 Research question: How is organizational integration achieved?

Table 22: Theoretical background: Organizational integration and transition strategies

Structuration theory emphasizes reciprocal interactions between organizational context and daily work activities and outcomes. *On the one hand*, organizational contexts determine activities and outcomes of organizational members as they "[...] realize institutional orders within their day-to-day actions" (Jarzabkowski, 2008, p. 622). For instance, Orlikowski (1996) shows how the introduction of new software changes organizational contexts in which individual activities are integrated. She demonstrates that employees rearrange daily work activities as the software demands additional electronic documentation. *On the other hand*, individuals' activities and outcomes shape the organizational context by consciously or unconsciously applying specific transition strategies (Jarzabkowski, 2008, p. 622; Giddens, 1979, Giddens, 1984; Jones & Karsten, 2008). Orlikowski (1996) indicates that changed work procedures results in the adaptation of coordination mechanisms at the organizational level and changes organizational contexts reciprocally. For instance, as electronic documentation has become available and accessible to managers, evaluation schemes have been redesigned at the organizational level to make use of available data from the software. For this, managers assess explored information provided via the software, develop indicators for the incentive scheme and integrate these indicators with the organization's structure. In conclusion, individuals change organizational contexts by applying transition strategies, such as changing structural elements like incentive schemes.

Thus, to better understand the organizational integration of organizational innovation communities, we must (1) clarify the influence of organizational contexts on innovation activities and outcomes in organizational innovation communities and (2) identify transitions strategies to form organizational contexts.

2.1 Organizational integration

Community researchers stress the importance of organizational integration, such as cultural and structural integration, for organizational communities to flourish (Kodama, 2005b, p. 896; Brailsford, 2001, p. 20; Kimble & Bourdon, 2008, p. 466). *First*, cultural integration comprises sets of norms, values, beliefs, attitudes, and procedures (Denison, 1984) that value the activities of organizational communities (Kimble & Bourdon, 2008, p. 466). Specifically, innovation activities are fostered by cultures that embrace curiosity, thinking outside the box, and risk taking while retaining open and informal communication styles (Smith, 1998; Capon et al., 1992; Gupta & Wilemon, 1990; Andriopoulos, 2001; Goll, Sambharya, & Tucci, 2001). *Second*, structural integration includes the amalgamation of administrative procedures, e.g., goal system integration, provision of resources, integration of performance appraisal systems, etc. (Venters & Wood, 2007, p. 360; Lee & Choi, 2003, p. 179; Brailsford, 2001, p. 25; Gassmann & von Zedtwitz, 1999). Although it is well-known that organizational integration supports community viability at large via adequate cultural and structural integration, so far insight concerning the influence of these context factors on innovation activities and outcomes remains rather limited and often speculative. This leads to the first research question: *How does organizational integration, by means of cultural and structural integration, influence the innovation activities and outcomes of organizational innovation communities?*

2.2 Transition strategies

Also, scholars in the field of community research emphasize that organizational innovation communities need to be increasingly integrated in organizational contexts as they mature (Rosenbaum & Shachaf, 2010). From a structuration perspective two major transition strategies to foster the integration of organizational innovation communities may be especially fruitful (Crowston et al., 2001; Jarzabkowski, 2008; Karsten, 1995). *First*, interactive transition strategies refer to direct interactions among individuals to mobilize support for cultural transitions, in the sense that norms, values, beliefs, attitudes, and procedures are reassessed and changed over time (Jarzabkowski, 2008, p. 629). *Second*, procedural transition strategies relate to adaptations of administrative procedures to achieve structural transitions, in the sense that goal systems, budgets, performance appraisal systems, etc., are adapted (Jarzabkowski, 2008, pp. 629–630). Even though these two transition strategies to foster organizational integration are highlighted in management research, to date very little is known on how to alter the integration of organizational innovation communities to trigger innovation activities and to attain innovation outcomes. This leads to the second research question: *How do transition strategies support the anchoring of organizational innovation communities within organizations?*

3 Research methods

Building on case study methods this part is an integral part of the design science approach guiding this dissertation. In the following it is explained how the case study approach feeds the design of the IT artifact. Briefly, data was complementary collected in two fundamentally different settings. First, in an early stage of the design science process the data of nine cases was collected in highly innovative organizations across various industries which have considerable experience with community innovation. Second, three pilot organizations were an additional source for data collection. Therefore data collection began at an early stage of the design science process, but accompanies one complete cycle of anchoring organizational innovation communities within these organizations. The data collected from a total of twelve organizations therefore gives 'snapshots' and longitudinal data of organizational integration. All organizations typically apply a wide range of innovation methods and possess versatile practical experiences about innovation-supportive contexts and how to shape them.

3.1 Sample and data collection

A multiple case study method is applied to answer the main research question (Eisenhardt, 1989; Yin, 2003). The **sample** includes twelve in-depth case studies from varying industry sectors. A major precondition for the case selection has been that they have considerable expertise in the application of innovation methods and designing organizational contexts for innovation pursuits. Additionally, all participating organizations have considerable experience with organizational innovation communities. While all organizations may be considered as top innovators, four companies are explicitly listed in well-known innovation rankings, such as 'The world's most innovative companies' (McGregor, 2006). Consequently, as innovation activities are integral components of these organizations, they are insightful sites for research concerning the phenomenon under study.

The sample is further split into **two sub-samples**. First, nine innovative organizations are identified that fulfill the above-mentioned requirements. One major objective to include these organizations in the sample stems from the need to gain insights on the role of organizational contexts before implementing the Open-I platform pilots. Additionally, these organizations are studied to increase generalizability and reduce the risk of selection bias concerning the pilot organizations by triangulating gathered data. Another risk stems from possible induction of results in the pilot studies as intensive collaboration has been established. To counteract this risk, the nine additional case studies deliver valuable insights concerning possible selection biases. Additionally, it is particularly important to mention that the nine organizations in this sub-sample cross different industry sectors and are of various sizes. Second, three organizations participate actively in the Open-I project. These organizations introduce and apply the developed Open-I platform. All these three organizations introduce the same platform with minor organization-specific adaptations and similar proceedings for introducing and managing are applied to ensure comparability. They are located within a realistic travel distance to ensure rich and recurrent qualitative data collection. The research project accompanies them closely and actively for a typical time period of three years. Particularly, joint research meetings as well as workshops within the organizations were instances of joint activities between researchers and practitioners.

Data collection in all twelve cases is mainly based on interviews (see table and further explanations on the next page).

Organization	Employees	Professions of interviewees	Interviews	Business
Organization 1	50,000–100,000	Innovation manager, strategic manager	9	Employment agency
Organization 2	15000–30,000	Innovation manager, strategic manager, project manager, human resource manager	8	Service provider
Organization 3	5,000–15,000	Innovation manager, project manager, strategic manager, human resource manager, executive officer	8	Accounting / Auditing
Organization 4	> 100,000	Innovation manager, strategic manager	2	Communication service
Organization 5	> 100,000	Innovation manager	2	Airline
Organization 6	50,000–100,000	Innovation manager, project manager, engineer	5	Automotive manufacturer
Organization 7	50,000–100,000	Innovation manager	2	Semiconductor
Organization 8	5,000–15,000	Project manager, innovation manager, mechanical engineer	3	High-tech manufacturer
Organization 9	< 5,000	Strategic manager, consultant	2	Innovation consulting
Organization 10	< 5,000	Strategic manager	1	IT solutions
Organization 11	< 5,000	Strategic manager, software engineer	2	IT services
Organization 12	< 5,000	Marketing manager, consultant	2	Innovation solutions

Table 23: Case study companies

To achieve high levels of credibility of interview results, the establishment of trustworthy relationships was a high priority, in order to support ongoing open and honest communication. Interviewees were explicitly asked to underline their statements with experiences and stories to reduce potential *post hoc* response bias (Golden, 1992). In-depth interviews, using a semi-structured interview guideline, lasted from 45 to 120 minutes, averaging 90 minutes. All interviewees held an academic degree (around 30 % held a PhD). To gain comprehensive insights, interviewees were selected from different departments, if possible. Usually, at least two interviews per case were conducted. With the exception of two cases, all interviews were audio recorded. Most often face-to-face interviews were conducted. However, in two exceptional cases telephone interviews were necessary due to geographic distance. For the first sub-sample of cases, data were enriched by extensive documentary analysis, including official firm reports and additional reports from independent sources. For the second sub-sample, interviews were conducted at two points in time to capture the longitudinal nature of the pilots. A first round of interviews was conducted before the Open-I platform was introduced along with the interviews in the first sub-sample. Additionally, at the end of pilot project another round of interviews was conducted to capture experiences and retrospective insights from the actual implementation process. Moreover, topics of major concern relate to experiences of varying stakeholders within the organization during the implementation of the Open-I platform and adds rich information to interpret collected qualitative data. Detailed notes were taken and additionally audio or video taped on many occasions of the collaboration process, if possible. This triangulation strategy is applied to control for potential biases inherent in qualitative research (Yin, 2003; Eisenhardt, 1989).

3.2   Data analysis

The considerable database of 46 interviews resulting in around 70 hours of interviews, countless hours of collaboration with pilot organizations, and more than 1,000 pages of transcripts (estimated) made it essential that data analysis procedures were rigorously applied. Finding and defining categories certainly has creative aspects (Langley, 1999, p. 691) or as Krippendorff (1980, p. 76) says "[h]ow categories are defined [...] is an art", but certain techniques can be followed to ensure quality of data analysis and adequacy of interpretation. To ensure rigor, Mayring's (2002) suggested five-step research procedure was followed.

Practically, analysis of data is split into five sequential steps: determination, explication and revision of categories, summative check, and interpretation. First, all transcribed data were screened, marked, and tagged, and preliminary categories were developed. Accordingly, transcripts, notes from meetings, and extensive documentation were analyzed following typical content analysis procedures (Mayring, 2002; Miles & Am Huberman, 1994; Ryan & Bernard, 2000). Second, preliminary categories were explicated in terms of clear definitions and descriptive examples from the data. These first definitions were discussed with other researchers to increase the objectivity of data analysis. However, in situations of confusion or contradictions interviewees were asked to clarify their meaning (Yin, 2003). Third, based on preliminary definitions and descriptive examples, the adequacy of all tags was reassessed and categories were revised to represent the data at hand more comprehensively. In particular, reliability was checked by coding data of the twelve in-depth case studies by three independent coders (Mayring, 2002). Fourth, developed categories were again checked in detail and then fed back to interviewees. A high degree of congruence indicates broad applicability and validity of results. Fifth, based on derived categories a taxonomy that captured the most important insights from the data set was derived, as presented in the findings section. The following screenshot shows data coding using Atlas.ti at an early stage of data analysis. On the right side tags and labels that refer to specific statements from the interviews on the left can be identified. In line with presented theoretical backgrounds, results concerning organizational integration are analyzed prior to transition strategies.

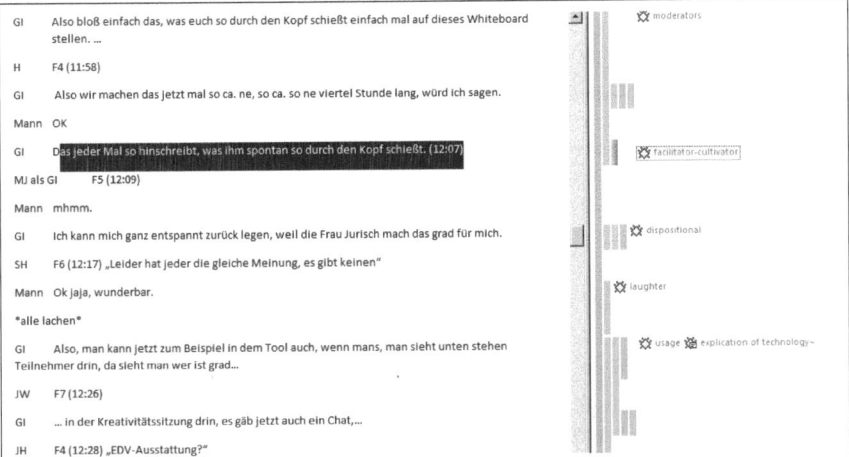

Figure 4: Example of using Atlas.ti for data analysis

4 Findings

The following section answers the first research question, concerning the influence of cultural and structural integration in organizational contexts, on distinct forms of innovation activities and outcomes. Thereby, this section gives Tom information concerning the influence of organizational contexts on innovation activities and outcomes and thus helps to solve the first part of his struggle, i.e. missing understanding concerning the influence of organizational contexts.

4.1 Influence of organizational integration

In sum, Tom learns that cultural and structural integration are of major importance for innovation activities and outcomes of organizational innovation communities. Specifically, he may be interested in the finding that *cultural integration* creates an innovative spirit, which triggers innovation activities and outcomes, by means of intensive and goal-directed communication among employees. Moreover, *structural integration* offers resources and aligns daily work activities to promote innovation activities and outcomes. Tom is provided with a taxonomy of four distinct types of organizational integration. Each type of organizational integration triggers distinct sets of innovation activities and produces divergent innovation outcomes. This information may be particularly meaningful to Tom as he may trigger those innovation activities and outcomes that are of strategic importance for his organization. The following section presents these four types of organizational integration in greater detail, including respective characteristics, innovation activities and outcomes. In the remainder of the section, overall characteristics of each type of organizational integration are described, typical innovation activities and how they unfolded in the cases displayed and respective innovation outcomes in terms of amount and timely occurrence presented. Numerous illustrations from the data are provided to enrich the derived categories with examples.

Dyadic integration combines the cultural integration (e.g., sets of norms, values, beliefs, etc.) and the structural integration of organizational innovation communities (e.g., incentive schemes, budgets, etc.). As a result of the combination of cultural and structural integration the data analysis reveals unique sets of 1) innovation activities and 2) outcomes. Before turning into the detailed presentation of findings, quotation 1.1.5 of table 25 illustrates how dyadic integration is administered in practice. In the example, it is shown that incentive schemes, e.g. (financial) reward for a created innovation, are combined with the cultural aspect. In this case, the direct superior manager hands over the financial reward, giving the employee recognition for his innovation activities. This activity clearly relates to innovation-oriented values of the organization. The subsequent table summarizes and visualizes key insights concerning dyadic integration.

Dyadic integration ignites distinct sets of *innovation activities*. The case analysis identifies two types of innovation activities emerging from dyadic integration: 1) the adaption of innovations to strategic objectives, i.e., cultural integration, and 2) pro-active seeking for innovation opportunities, i.e., structural integration (detailed explanations follow below the table on the next page).

Dyadic integration combines cultural [middle] and structural [left] integration for innovation pursuits [right]

 First, **findings** show that *innovation activities* of community members are characterized by constantly seeking for innovations. Moreover, activities include adaptation of these innovations to strategic objectives

 Second, **findings** show that *innovation outcomes* under conditions of dyadic integration emerge frequently and numerously.
In sum, dyadic integration [lower left] leads to constant seeking for strategically relevant innovations [upper middle]. This leads to recurrently high amounts of innovations [lower right]

 Tom **learns** about the influence of dyadic integration on innovation activities and outcomes of community members

 Project example: One Open-I pilot organization has decided to strategically implement organizational innovation communities as a means to further innovation development. Three crucial incidents have led to dyadic integration. First, top and innovation managers have convinced middle managers that communities have the potential to unleash creativity in the pursuit of innovation and that they also have the ambition to consequently establish such communities. In so doing, middle managers have become aware of advantages of such communities (cultural integration). Moreover, middle managers have identified strategic innovation challenges to profit most from these communities. As the identified innovation challenges clearly related to agreed individual objectives of these managers, the community is structurally integrated. Third, middle managers give employees the time to work on the identified innovation challenges but also emphasize the importance during lunch or weekly meetings or other personal interactions. In sum, organizational innovation communities are culturally integrated, in terms of top management emphasize, and structurally, in terms of individual objectives and time given to employees.

Table 24: Dyadic integration: Definition, findings, learning, and project example

First, community members profit from cultural integration by (i) being able to rely on open and informal communication styles and (ii) by being provided with a clearly explicated frame for their innovation activities. This leads to innovation activities being characterized by a creative and proactive *adaption of innovations* to strategic objectives. Reaching for the compatibility of innovation and strategic objectives, community members experience social reinforcement in their innovation activities, in the sense that peers as well as leaders embrace attempts to develop innovations (for illustration see quotations 1.1.1 and 1.1.2). The examples demonstrate that employees are able to communicate across the boundaries of the organization, for instance departmental or hierarchical boundaries. It also becomes clear that this form of cultural integration fosters 'out of the box thinking'. Additionally, as employees are asked about their 'own' innovation projects in goal setting talks, innovation activities are even furthered and aligned to strategic objectives.

Second, structural integration leads community members to *constantly seek innovations*. The case analysis reveals that community members under conditions of structural integration are cognizant to seize innovation opportunities when they emerge from daily work procedures. For instance, quotation 1.1.2 shows that a multitude of methods are applied to support every-day innovation activities. These include the provision of IT-based social networks, community-oriented workshops, provision of resources, controlling instruments and community-based incentive schemes, etc. (for illustration see quotations 1.1.3 or 1.1.4). In sum, findings show that structural and cultural integration are two major and equally important means to foster innovation activities (for illustration see quotations 1.1.6 and 1.1.7). For instance, in organization 7 cultural and structural aspects are seen as two sides of *one* medal that need to be integrated.

No.	Quotations concerning dyadic integration
Innovation activities	
1.1.1	Organization 7: 'Cooperating with other departments without hesitance is daily business. [...] Employees regularly start to think outside the box which leads to ideas for new products.'
1.1.2	Organization 9: 'Additionally, we provide monetary / extrinsic incentives and things like specific titles. [...] However, employees' own innovation projects are always included in goal setting talks.'
1.1.3	Organization 7: 'Firm principles are printed on business identification card of every employee. They summon to think lateral and not to keep ideas for themselves, but to discuss them in groups.'
1.1.4	Organization 7: 'There is special training to give new employees an understanding of the culture, to show how business processes are executed and how to apply methods.'
1.1.5	Organization 3: 'This is how we do it: If someone earned a reward for an innovation, we leave it to the superior to hand over the reward. We think it is very important that employees get recognition along with financial reward.'
1.1.6	Organization 7: 'On the one hand, there is the creative side and on the other the controlled side. For instance, we have clear guidelines concerning how to manage our main businesses.'
1.1.7	Organization 7: 'The two major forces are firstly attempts to develop and maintain an innovative culture [...] and secondly concerns about how to simplify and accelerate methods, tools and processes to strengthen innovation endeavors.'
Innovation outcomes	
1.2.1	Organization 9: 'Our company is unique, as our innovative philosophy is deeply rooted in our firm principles, but also in the lived culture. [...] For instance, we consciously maintain creative chaos.'
1.2.2	Organization 9: 'What we developed appears to me to be a radical innovation. Besides, what is available in the internet for us it was something completely new. [...] However, the next such innovation is already about to be executed.'
1.2.3	Organization 7: '[...] Our history shows that we cannibalize our own products, as we come up with new stuff that questions or replaces older solutions.'

Table 25: Quotations concerning dyadic integration

As Tom now knows how employees react on dyadic integration in terms of activities, he is curious about the **innovation outcomes** created by the dyadic integration of organizational innovation communities. The cases reveal that a self-reinforcing system combining cultural and structural integration leads to innovation outcomes characterized by high amounts of innovation[13] developed recurrently: Based on strategically important innovations, community members experience cultural support, e.g., by means of social support from peers and leaders, in a first step. As a consequence of cultural reinforcement, they are eager to achieve *high amounts* of strategically important innovations. These

---

[13]In numbers, each participating employee developed one innovation per year based on around 20 to 30 initial ideas.

innovations are supported by structural integration, in terms of (i) resources given to follow-up innovation ideas and (ii) incentives for community members working on innovation concepts. Organization 7 shows that these high amounts of innovations are created within a consciously maintained 'creative chaos' that reflects the organizations culture to create high numbers of innovations. In a second step, results show that community members attempt to *produce innovations recurrently* as they experience recognition and support not only socially, but also structurally. For instance, during goal setting talks, i.e. a structural element, employees are explicitly asked for their 'own' innovation projects (quote 1.1.2). In sum dyadic integration, i.e. cultural and structural integration, results in innovation outcomes characterized by high amounts of innovations developed recurrently (for illustrations see quotations 1.2.2 and 1.2.3.).

	Cultural integration refers to organizational contexts in which organizational innovation communities are culturally [left] but not structurally integrated in the pursuit of innovation [right]. For instance, CEOs honor highly performing community members personally under conditions of cultural integration	
	First, **findings** show that *innovation activities* of community members under cultural integration are characterized by adaptation of innovation concepts to strategic objectives of the organization if triggered by leaders	
	Second, **findings** additionally demonstrate that numerous innovations are created (*innovation outcomes*), if triggered In sum, cultural integration [left] results in community members adapting innovations to strategic objectives if triggered [upper middle], creating numerous strategically relevant innovations [lower right; the diamond in the light bulb may be associated with *strategic relevance*]	
	Tom **learns** about the impact of cultural integration on innovation activities and innovation outcomes	
	Project example: In an early stage of a pilot study, the consortium of researchers and practitioners is taken in by the somewhat naïve aspiration that if community platforms are provided, employees will 'automatically' engage. However, due to missing initial engagement an open discussion round (this session could also be regarded as an emergency meeting) was set up by the responsible managers. They emphasized that the project has high top management visibility and they encouraged everyone to participate more actively. Gaining awareness of the importance of the project, employees engaged in the community every time they were asked to. Consequently, this example shows that organizational innovation communities are culturally integrated as managers put (social) pressure on affected employees.	

Table 26: Cultural integration: Definition, findings, learning and project example

Cultural integration displays situations in which organizational innovation communities are culturally, but not structurally integrated. For instance, in organization 8 created innovations are honored by the CEO's recognition (cultural integration), but incentives are not provided (missing structural integration). Consequently, due to missing structural integration, innovation activities and outcomes are not integrated into incentive schemes, controlling instruments, etc. However, due to cultural integration, social support and pressure drives community members to participate in organizational innovation communities. As a result of cultural integration and a lack of structural integration the data analysis reveals unique sets of 1) innovation activities and 2) outcomes. Before turning to these aspects, the following table encapsulates key findings.

Cultural integration of organizational innovation communities leads to specific *innovation activities*. As shown under the conditions of dyadic integration, innovation activities that typically emerge under conditions of cultural integration refer to the *adaption of innovations* to strategic objectives. For instance, quotation 2.1.1 of table 3 shows that innovations with an impact on strategic objectives are explicitly honored. Moreover, organization 4 shows that managers communicate the areas they want to further their innovation activities. Consequently innovation activities concentrate on these areas. However, in contrast to dyadic integration, community members *do not constantly seek for innovations* under conditions of solely cultural integration, as their efforts are not structurally supported, e.g., by means of rewards of any kind or accessibility of resources to market innovations, etc. (for illustration see quotation 2.1.2). Also in organization 4 it became clear that employees working on innovation tasks were seen as a 'slack resources' and hence had to help out in other projects as soon as they become urgent. Consequently, innovation activities are interrupted as they are not structurally integrated, in this case by means of time.

Tom asks himself to what extent cultural integration constrains *innovation outcomes* or if the negative impact of missing structural integration remains negligible as community members are still culturally integrated. The data analysis speaks a clear language: cultural integration leads to contradictory work situations for community members. Top managers still support innovation activities and are curious about outcomes. However community members do not have the resources, for instance in terms of time or budget, to follow-up their innovation activities to produce innovation outcomes regularly. As a result, innovations are still developed but do not evolve recurrently. On the one hand, structures 'over-emphasize' daily tasks to be fulfilled. For instance, direct supervisors (or in other terms team leaders) penalize innovation activities as they aim at achieving different and pre-defined goals. This is illustrated by statement 2.2.1 showing that employees are permanently occupied by daily tasks and do not have the time to engage in organizational innovation communities. Consequently, the innovation potential of community members is often even hindered. On the other hand, organizational cultures advocate innovation activities in organizational innovation communities. Social pressure, in terms of intensive interactions and persuasion, delivers a major impetus for community members to participate in innovation activities and ignites considerable efforts to produce high amounts of innovations. Consequently, *innovation outcomes are not generated recurrently*, but occur if triggered through social pressure. Quotation 2.2.2 delivers one example of how social pressure can function: Within specific meetings, innovation managers clearly put pressure on community members which did not participate enough or as promised.

In sum cultural integration results in innovation outcomes characterized by high amounts of innovations if triggered by top managers, i.e. exerting social pressure. For Tom, this means for Tom that innovation activities and outcomes have to be pro-actively triggered by himself or his colleagues but do not emerge from community members themselves. The good news is that as soon as Tom triggers innovation activities and outcomes, in most cases he will be provided with innovation results.

No.	Quotations concerning cultural integration
Innovation activities	
2.1.1	Organization 8: 'Employees who solved an important issue – e.g., manufacturing issues – are honored. For instance, this includes an invitation to the CEO every two years, an award, a ceremony, etc.'
2.1.2	Organization 4: 'Employees working on innovation projects are typically seen as slack resources. They are used to balance out peaks in labor demand. Consequently, innovation projects often die or at least suffer from a lack of work force.'
2.1.3	Organization 4: 'Innovation mostly happens within our major strategic objectives, as we communicate: 'These are the areas we want to facilitate and strengthen innovation'.'
2.1.4	Organization 5: 'We have a defined process, but honestly speaking, this process is more like arrows and boxes than lived culture. We currently work to bring it to life.'
2.1.5	Organization 3: 'The CEO participated once. Consequently, employees participated heavily.'
Innovation outcomes	
2.2.1	Organization 6: 'To recurrently participate, employees are way too much occupied by daily working procedures.'
2.2.2	Organization 4: 'We try to increase priorities for our community members by setting up specific meetings […]. Within these meetings we also educate our members, as we clearly explain members [with low participation] that they did not perform and that they should do better.'
2.2.3	Organization 4: 'The one with the highest benefit from the innovation project organizes the project and is responsible to motivate employees to participate.'
2.2.4	Organization 4: 'Mostly, innovations include principles for new business areas or new business models and may be seen as radical innovations.'
2.2.5	Organization 3: 'The CEO participated once. Consequently, employees participated heavily. A lot of innovations resulted from this session. Some of the topics were known for a long time, but were never implemented.'

Table 27: Quotations concerning cultural integration

Having learnt how cultural integration impacts innovation activities and outcomes, Tom now speculates about how **structural integration** influences organizational innovation communities. Structural integration unfolds as organizational innovation communities are not culturally, but structurally integrated. Particularly, by virtue of missing cultural integration, innovation activities and outcomes are not part of organizational norms, values, beliefs, attitudes and procedures and hence are not socially reinforced. Particularly, innovation activities and outcomes are mostly driven by incentive schemes, controlling instruments, etc. Consequently, structural integration induces unique sets of 1) innovation activities and 2) outcomes. In the subsequent table key findings referring to structural integration are displayed.

Structural integration relates to structures that integrate community-related activities organizationally [left] for innovation pursuits [right]. For example, the integration of community-related activities in goal systems and incentive schemes or provision of budget are means of structural integration

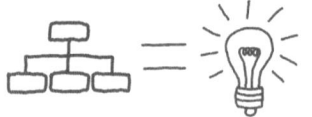

 First, **findings** suggest that *innovation activities* are characterized by community members, pro-actively searching for innovations

 Second, **findings** additionally show that innovations are created recurrently as they are part of the organizations' structures (*innovation outcomes*)
Consequently, structural integration [left] triggers community members to pro-actively search for innovations [upper middle], creating a recurrent flow of innovations [lower right]

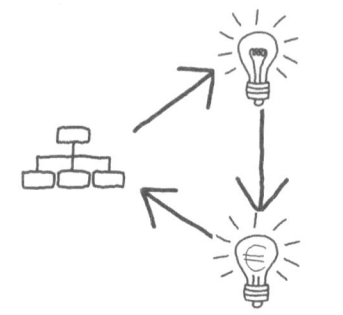

 Tom **learns** that structural integration influences innovation activities and triggers innovation outcomes in a different manner compared to dyadic and structural integration

 Project example: During the implementation of an organizational innovation community in a pilot organization, another form of integration is observed: structural integration. As participation of employees in the organizational innovation community remains below expectations, it is decided to integrate the community structurally. The responsible manager sends out invitations to employees to engage in creativity sessions on the Open-I platform using the outlook calendar function. The reason for this step has been that employees heavily rely on their outlook calendars as their major tool to organize daily work: 'Everything that is in the calendar is prioritized and gets done'.[14] In other words, the outlook calendar invitations are a means to give employees the time to work in the organizational innovation community, i.e. to integrate these community members structurally.

Table 28:Structural integration: Definition, findings, learning, and project example

Innovation activities under conditions of structural integration are best described by community members recurrently seeking for innovation opportunities. Similar to conditions of dyadic integration, innovation activities are structurally supported by means of incentive schemes, resources, etc. (for illustration see quotation 3.1.1 of table 29). The quotation shows that structural integration leads employees to search for innovations 'all through the year', i.e. recurrently. Moreover, central budgets for innovation deliver resources for innovation activities which would otherwise be canceled (for illustration see quotation 3.1.2). The quotation shows that the provision of a central budget for innovation activities is crucial for fostering organizational innovation communities. It clearly indicates that innovation activities would not be agreed upon by direct superiors, leaving smaller budget expenditure for other activities. Consequently, community members *pro-actively seek for innovation opportunities* recurrently. Despite the positive influence of structural integration on innovation activities, a lack of cultural integration has a negative effect on these activities. Admittedly, two major effects of lacking cultural integration are found. Community members can neither rely on open and informal communication nor are they provided with a clear frame within which to align their innovation activities. For instance, quote 3.1.5 emphasizes that structural integration may ensure a defined number of innovations be created within a given time span,

---

[14] It is important to mention that the manager is not a direct supervisor of the invited employees.

but these innovations do not necessarily have strategic relevance. The data analysis reveals that such circumstances *undermine effective adaption of innovations to strategic objectives* and prohibit social support for innovation activities. Quotation 3.1.2 and 3.1.5 show this effect.

Innovation outcomes under conditions of structural integration are characterized by low amounts of innovations but they emerge recurrently. The data analysis reveals that structural integration delivers constant motivation for community members to produce innovations. For instance, quotation 3.2.2 shows that the inclusion of innovation objectives in job descriptions fosters constant flows of innovations. Hence, community members *develop innovations recurrently*, along other obligations in their daily work. Despite this finding, missing social support seems to lower community members' efforts to develop high amounts of innovations. The data indicate that community members stop innovating as soon as they produce a particular number of innovations, as defined by the job description or in goal setting talks, i.e. as structurally defined. Quote 3.1.5 gives some insights on the functioning of defining a number of innovations to be created in a given time span. Tom may be particularly interested in the finding that employees submit projects they would have done anyway or develop 'alibi projects' to achieve these numbers. Quotation 3.2.1 illustrates that missing social support reduces outcomes in organizational innovation communities. In this case the number of employees engaging in organizational innovation communities remained too low to create situations of social support and pressure, i.e. cultural integration. Hence, the amount of developed innovations remains at a low level.

No.	Quotations concerning structural integration
Innovation activities	
3.1.1	Organization 8: 'Innovation objectives are included in employment contracts. [...] Consequently, we think about innovation all through the year.'
3.1.2	Organization 8: 'Without this budget? We would have to ask our supervisor if he was willing to invest. However, no supervisor would give money from his budget without the need to do so. [...] Therefore the central budget is a good mechanism to facilitate short-term innovations.'
3.1.3	Organization 8: 'Employees are organized in projects, what means that they are not free to spent working time on innovations as they wish. For such cases [in which employees have innovative ideas] we installed a centrally organized innovation budget.'
3.1.4	Organization 4: 'Employees need to have some basic impetus to participate. This includes not only financial incentives, but more important training.'
3.1.5	Organization 1: 'Managers of every department have to produce 3 innovation project each year. [...] However, some department managers include what they would do anyway and thus the influence on fostering innovativeness is 'zero' while others do an 'alibi-project'.[...] So, the pure amount of innovations we develop remains constant each year or even each quarterly, but they are not necessarily strategically meaningful.'
Innovation outcomes	
3.2.1	Organization 6: 'However, corporate innovation communities only function if enough employees participate actively. There is a critical mass for social processes to function. This has, in my opinion, not yet been reached in our system.'
3.2.2	Organization 8: 'Innovation objectives are included in employment contracts. During appraisal interviews, employees are encouraged to articulate innovation proposals. Consequently, we think about innovation all through the year.'
3.2.3	Organization 8: 'We established a quite tough process to push innovations to market. We govern innovations through quality gates.'
3.2.4	Organization 8: 'Innovations are task-related but to an equal extent out of scope.'
3.2.5	Organization 4: 'Additional to knowledge gains and motivational aspects, trainings would additionally show management commitment which is not obvious today.'

Table 29: Quotations concerning structural integration

No integration describes situations in which organizational innovation communities are neither culturally nor structurally integrated. In this case, neither are innovation activities and outcomes part of organizational norms, values and beliefs nor are they reinforced by incentive schemes, controlling instruments, etc. Consequently, due to a lack of cultural and structural integration 1) strongly limited innovation activities occur and 2) only a few innovation outcomes evolve. A summary of 'no integration' is provided in the subsequent table.

 No integration describes organizational contexts which neither integrate organizational innovation communities culturally nor structurally [left, middle and right]

 First, findings demonstrate that *innovation activities* remain sparse under conditions of 'no integration' as daily tasks have to be fulfilled

 Second, findings additionally show that *innovation outcomes* remain limited as employees lack a basic impetus to get active in organizational innovation communities

 Tom learns that organizational innovation communities do not emerge all by themselves but need to be supported culturally, structurally or both to achieve innovation activities and outcomes

Table 30:No integration: Definition, findings, and learning

First, *innovation activities* under such conditions remain limited in terms of community members seeking for innovations and adapting them to strategic objectives. For instance, innovation activities are not or not effectively supported structurally by means of incentive schemes, resources, etc. For illustration, quotation 4.1.1 of table 31 shows that performance appraisals guide employees' activities. As innovation activities are not connected to compensation in any form, employees do not seek for innovations constantly. Moreover, under conditions of no integration, innovation activities in organizational innovation communities are not supported socially, in terms of the organization's culture as it is illustrated by quotation 4.2.2.

Second, *innovation outcomes* under conditions of no integration are characterized by low amounts of innovations which emerge occasionally, if at all. The data analysis reveals that the lack of both structural and cultural integration results in a motivation deficit. Quotations 4.1.1 and 4.1.2 illustrate this condition of missing structural and cultural integration. Further, quotation 4.2.2 displays how employees react to this lack of organizational integration: they do not have the impetus to develop innovations. Moreover, missing cultural and structural integration limits overall innovation outcomes as quotation 4.2.1 shows.

No.	Quotation concerning neither cultural nor structural integration
Innovation activities	
4.1.1	Organization 3: 'Performance appraisal is a central instrument to guide employees in our firm. [...] Innovation topics are enlisted in terms of qualitative objectives. Employees shall develop innovations, admittedly innovation is not linked to compensations in any form.'
4.1.2	Organization 3: 'Innovations are not honored specifically. [...] Innovation is specified in our firm objectives; however it is not consequently followed-up. [...] I have to admit that innovation-related topics are not communicated with any emphasize in our firm.'
Innovation outcomes	
4.2.1	Organization 3: 'When it comes to innovation, we have to admit that we are not the fastest.'
4.2.2	Organization 2: 'As long as employees do not have any incentive to participate they will not develop innovations. [...] By incentive, I think of financial rewards, career development, etc., but also of attention from peers and team leaders.'

Table 31: Quotations concerning neither cultural nor structural integration

4.2   Transition strategies

In the previous section, Tom has learned how organizational integration, in terms of cultural and structural integration, brings about distinct sets of innovation activities and outcomes. He learns that if community members, constantly seeking for high amounts of strategically relevant innovations, are the main goal, he has to dyadicly integrate organizational innovation communities. However, if he wants high amounts of strategically relevant innovations 'on demand' he would probably aim at integrating organizational innovation communities culturally. Alternatively, if the major aim is to have employees pro-actively and recurrently searching and developing innovations, then structural integration is more appropriate. Having identified these distinct influences which support or hinder the integration of organizational innovation communities, the main focus shifts towards the second research question: *How do transition strategies support anchoring of organizational innovation communities within organizations?* This question is especially important to Tom as he has to create organizational contexts to realize relevant innovation activities and outcomes.

Findings show three distinct transition strategies that describe procedures to increase organizational integration in terms of cultural and/or structural integration. *Initiating strategies* refer to the emergence of organizational innovation communities. Applying *negotiation strategies*, organizational innovation communities are either structurally or culturally integrated. *Narration strategies* lead to further integration resulting in dyadic integration. The identified transition strategies are critical means to alter organizational integration. In the following section, each transition strategy is explained in greater detail and examples from the case studies are provided.

Tom might be interested in the way organizational innovation communities emerge from employees' daily work issues to facilitate this process consciously in the future. While analyzing the data two major means of **initiating strategies** that trigger the emergence of organizational innovation communities are identified. A summary of key findings is presented in the following table.

	Initiating strategy describes the facilitation of emerging organizational innovation communities
	Findings suggest that unsolved innovation challenges deliver a basic impetus for community members to engage [left]. Emotional involvement concerning these challenges [upper middle] and curiosity leading to engagement [lower middle] potentially creates conditions under which organizational innovation communities emerge 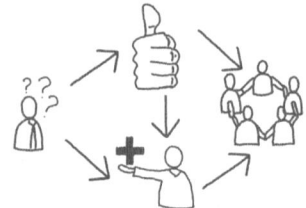
	Tom **learns** that particular circumstances fosters the emergence and first appearance of organizational innovation communities
	Project example: Reporting from an early stage of implementing the organizational innovation community, one crucial strategy concerning the emergence of organizational innovation communities became visible. Despite a lack of employee engagement in the organizational innovation community, in one meeting it became clear that several employees were regularly confronted with complaints concerning a particular service they provide. The complaints were especially incriminating as these employees were well aware of the issue. They had aimed at changing this service before individually but were not successful. However, together they felt like they were able to significantly improve the service, but also to reduce the number of complaints. Jointly, they decided to take action and to work out solutions to improve the service. At present, the solution is partly implemented.

Table 32: Initiating transition strategy: Definition, findings, learning, and project example

First, analyzing the data reveals that perceptions of present work situations provide a crucial impetus for organizational innovation communities to emerge. Particularly, important but unsolved innovation-related issues ignite the emotional involvement of employees. Then, curiosity concerning these kinds of issues potentially leads to attempts to solve them. Finally, a positive perspective to solve such issues enhances the probability of organizational innovation communities evolving (for illustration see quotations 5.1.2 and 5.1.3 of table 6).

No.	Quotations concerning initiation strategy
Preconditions	
5.1.1	Organization 9: 'Basically, it works like this: Someone is thinking creatively, articulates an innovation idea and invites other employees to participate. [...] Everyone has his/her own reasons why they engage in innovation development.'
5.1.2	Organization 2: 'Application of software [topic changed due to confidentiality] was an issue for a long time. Now, we want to address it. [...] Something has to happen now.'
5.1.3	Organization 2: 'I think, there is a lot of potential around the topic.'
Transitions	
5.2.1	Organization 5: 'Initially, our department was not appointed by our company [...], but we developed from an own idea: we developed our business field independently and we term it innovation.'
5.2.2	Organization 5: 'As long as our figures are acceptable and we do not crash a product, we are free to do what we want.'

Table 33: Quotations concerning initiation strategy

Second, building on fertile seeding grounds, employees anchor organizational innovation communities in organizational contexts in specific manners. For instance, as community members advance innovations, it becomes hard for top management to disapprove of their application. Moreover, as community members invest considerable effort in advancing an innovation, they still foster their attempts even though the benefits do not outweigh their effort. Hence, organizational innovation communities are anchored in organizational realms as community member advance innovations considerably and invest considerable effort. However, this type of anchoring is not substantial and may vanish as soon as the major issue of concern is solved.

Findings indicate that **negotiating strategies** establish either structural or cultural integration based on community members' motivation to change unsatisfying innovation activities or outcomes by identifying critical organizational context factors. Changing these critical factors ultimately leads to the adaption of organizational contexts as shown in the following. Before turning to an in-depth analysis of negotiating strategies, key insights are summarized in the following table.

 Applying **negotiating strategies** establishes either cultural or structural integration based on unsatisfying innovation outcomes

 Findings demonstrate that unsatisfying innovation outcomes [left] motivate community members to pro-actively search for solutions to overcome critical barriers hindering innovation outcomes [middle]. Community members adapt organizational contexts, either cultural or structural [lower and upper right], that promise to have the greatest impact on innovation outcomes with the least effort

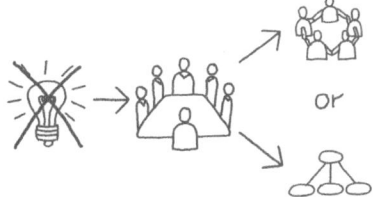

 Tom **learns** that community members are able to identify crucial organizational barriers hindering innovation outcomes, either culturally or structurally

 Project example: At one point, innovation activities nearly came to a complete halt on the community platform. An emergency meeting was called, which revealed that community members were still motivated to engage in innovation activities. However, during the meeting it became clear that time pressures and workload inhibited sufficient engagement in the community. Also, it was mentioned that they feared that peers would label their engagement as 'leisure entertainment'. However, discussing their experiences and concerns they found ways of resolving them. For instance, community members asked for 'homework' as a structural element to legitimate their engagement in front of their peers.

Table 34: Negotiating transition strategy: Definition, findings, learning, and project example

First, unsatisfying innovation activities – for example, in terms of participation and innovation outcomes resulting from a lack of cultural and structural integration (as described above) – motivate community members to actively adapt the organizational context. For instance, in quotation 6.1.2 in table 35, an interviewee clearly articulates issues concerning the innovation activities and outcomes of the community, while at the same time emphasizing her motivation to retain community efforts. Second, community members identify central barriers limiting their innovation activities and outcomes (for illustration see quotation 6.1.3.). Based on the invitation to honestly discuss pressing concerns hindering innovation activities, community members open up to find solutions to integrate organizational innovation communities in a more sophisticated way. Third, they adapt those cultural or structural context factors that need least effort but offer most improvement (for illustration see quotation 6.2.3). Within organization 3 structural elements seemed difficult to change, so community members decide to apply a cultural element: Community members decided that one member may give 'homework', i.e. something that every community member has to prepare for the next discussion. While identifying crucial and easily changeable context factors, cultural integration is achieved as following: First, community members interactively develop sets of community norms and values which are based on the organization's culture but differ in important aspects. For instance, communication should be built on trust and should ignore political aspects. Furthermore, the data show that a community member following up agreed norms, etc., is crucial. Hence, the community's culture exerts social pressure among community members to actively participate and produce valuable outcomes (for an illustration of this see quotation 6.2.1). Clearly, they know that one person pushing the community is needed, as quote 6.2.1 shows. Structural integration is administered by aligning community activities with daily tasks. Reframing and interpreting set goals and objectives are an observed strategy community members may apply. For instance, quotation 6.2.2 shows that community members asked for 'homework' to better include community activities in their goal system.

No.	Quotation concerning negotiation strategy
Preconditions	
6.1.1	Organization 7: 'E.g., a myth concerning the founder's challenging work attitude and a mastered company-wide strategic realignment nurtured to maintain cultural support for innovation activities.'
6.1.2	Organization 3: 'At the moment, I do not like the way it [organizational innovation community] is working and you do not like it either. Still, the project is very important for me.'
6.1.3	Organization 3: 'So, I want to have an honest conversation about the most pressing concerns. […] Based on these experiences, I would like to advance and discuss with you, how we should move on in this project.'
Transitions	
6.2.1	Organization 2: 'At least one person is needed that pushes activities in communities. This person must definitely not be a leader from top management, but should be capable to put pressure on time frames, etc.'
6.2.2	Organization 3: 'I was pretty impressed that the community members asked for homework. But we tried it. Having homework was important to employees as they were able to justify participation that was not directly linked to their job descriptions.'
6.2.3	Organization 3: 'I do not think that we can change something concerning the incentive system. We would have to convince the top management, discuss changes with the work council, etc. This may take some years. […] The idea of homework sounds appealing, as we can apply this instantly without asking any top management staff or other stakeholders.'

Table 35: Quotations concerning negotiation strategy

The data analysis reveals that the communication of innovation outcomes in form of **narrations** plays a key role in integrating organizational innovation communities structurally and culturally. The following table condenses key findings concerning the transition strategy of narration.

	Narration strategies achieve cultural and structural integration of organizational innovation communities. They may be understood as success stories that are supported by a high percentage of community members
	Findings show that success stories in the form of narrations are an effective strategy to create curiosity among top managers and employees [lower left]. Additionally, achieved results affirm the effectiveness of organizational innovation communities for innovation development [upper left]. Based on evidence in the form of narrations, top managers agree to integrate these communities even more [middle], i.e. either by adding cultural [upper right] or structural support [lower right] 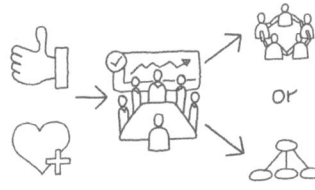
	Tom **learns** that narrations about successes achieved through community-based innovation activities are a crucial cornerstone to convince top managers to integrate organizational innovation communities dyadically
	Project example: Initial pilots within the project Open-I in all three organizations ended with a presentation of achieved results. In one case, innovation concepts were presented in front of the whole executive board. All board members were fascinated about the achieved results, so the CEO finally decided to set up innovation projects for half of all concepts presented. The other concepts were put on a waiting list as resources for more implementations were not available. The organization is currently screening platforms that provide key features of the Open-I platform to implement organizational innovation communities long term

Table 36:Narration transition strategy: Definition, findings, learning, and project example

First, community members identify examples that have the potential to show the effectiveness of community-based innovation development. These examples should have a positive emotional meaning to a high percentage of community members as illustrated in quotation 7.1.1 in table 8. Also, they have to be available in a vivid, rich and comprehensible manner (for illustration see quotation 7.1.2) in order to be easily communicated to top management (for illustration see quotation 7.1.3).

While communicating success stories to top managers and other interested employees, organizational contexts are towards dyadic integration. First, stories of successful innovation activities and outcomes increase perceived importance among top managers and employees at large, which anchors organizational innovation communities. In quotations 7.2.2 and 7.2.4 the interviewees emphasize that top managers get excited and want to take an active role within the community. Consequently, sets of attitudes, norms, and values are adapted and ultimately lead to cultural integration. Second, by their retrospective nature, stories are able to undeniably affirm the positive impact of organizational innovation communities on overall organizational innovativeness. In quotation 7.2.2 the interviewee states that half of the developed innovation concepts are under development, increasing the organization's innovation implementation drastically. For this, resources are provided to develop these innovation concepts. Quotation 7.2.5 shows that structural rearrangements follow the narration strategy to provide needed resources. Consequently, top management supports the integration of organizational innovation communities in a structural way, i.e. by providing budgets or other resources. In sum, narration strategies lead to structural and cultural integration, i.e., dyadic integration.

No.	Quotations concerning narration strategy
Preconditions	
7.1.1	Organization 2: 'From my experiences, it is vital to ignite a 'bottom-up process'. During this process, a network for innovation ideas emerges which is driven by employees breadthways of the firm. Later, innovations are presented to top management by a group of employees. However, at this time all involved employees are already convinced about the innovation.'
7.1.2	Organization 2: 'It will take some time until we will have achieved the goal to fully integrate the innovation community. We are now in a stage in which we have to convince through single projects to create a 'pull-effect'. [...] Until then, we still need a sponsor in a central position to develop wished results.'
7.1.3	Organization 3: 'We have to talk the language of our top management. [...] The way they are communicating heavily relies on business concepts.'
Transitions	
7.2.1	Organization 7: 'We are trying to nurture an innovation culture. For instance, everyone knows stories about the company founder's challenging and creative work attitude.'
7.2.2	Organization 2: 'The CEO was very fascinated by our results. Actually, he wants around half of the innovations we proposed implemented. That is unbelievable. In six months he insists on having a follow-up meeting to check whether the innovations have been implemented.'
7.2.3	Organization 2: 'I understand this idea and I think it is very interesting. [...] Actually, this is THE idea of the century.'
7.2.4	Organization 3: 'In the meantime [after applying the story strategy], we have eight top to middle managers who want to go through our process. Actually, this is a problem because we never expected so much enthusiasm and we have to find a solution for organizing this demand, we never planned for.'
7.2.5	Organization 2: 'After our presentation it turned out that we will be restructured to have the needed capacities, especially concerning work force, to bring the innovation concept to market.'

Table 37: Quotations concerning narration strategy

5 Discussion

An organization ready to implement organizational innovation communities might not be aware in detail of the role that different forms of organizational integration play in community members' innovation activities and resulting innovation outcomes. An organization that has already implemented organizational innovation communities might recognize that the existing integration of the community does not always lead to the intended results. Therefore, organizations need to learn about possible transition strategies, which allow the anchoring of organizational innovation communities in their preferred way.

By analyzing the antecedents and transitions of organizational innovation communities, a more nuanced understanding of the mechanisms driving the implementation and anchoring of organizational innovation communities is achieved. First, the empirical evidence of this study shows how organizational contexts foster innovation activities and outcomes. Second, findings show how transition strategies anchor organizational innovation communities in these contexts. Combining the findings of 'organizational integration' with the findings of 'transition strategies' allows the derivation of a taxonomy of organizational innovation communities. By doing this, this study adds to previous research in the field of community research and provides innovation management research with an interesting, yet still understudied perspective of analyzing two important determinants of the success of organizational innovation communities.

A **taxonomy** of four types of organizational integration – i.e. dyadic integration, cultural integration, structural integration, and no integration – results in a distinct set of innovation activities and outcomes (see figure 5). Depending on the type of organizational integration, different innovation activities are achieved by members of organizational innovation communities. These include the attempts of community members to adapt innovations to the strategic objectives of the organization as well as to pro-active seek innovation opportunities. By doing so, innovation activities result in distinct innovation outcomes which can be distinguished by 1) the amount of developed innovations and 2) whether they occur recurrently or occasionally. Having said this, findings also show that three distinct transition strategies – i.e. the initiation strategy, negotiation strategy, and narration strategy – allow the organizational innovation community to move from one type of organizational integration to another. It is argued that the initiation strategy leads to the emergence of organizational innovation communities without structural or cultural integration. The negotiation strategy integrates organizational innovation communities either structurally or culturally. This is done by grafting on unsatisfying innovation activities and outcomes as an impetus to identify crucial organizational context factors and adapt them. The narration strategy leads to the highest extent of integration as it leads to dyadic integration, i.e., cultural as well as structural integration. We can observe, therefore, four distinct types of organizational innovation communities and transition strategies for anchoring them. In sum, this study reveals (i) new insights concerning the role of organizational integration and (ii) transition strategies to strengthen organizational integration.

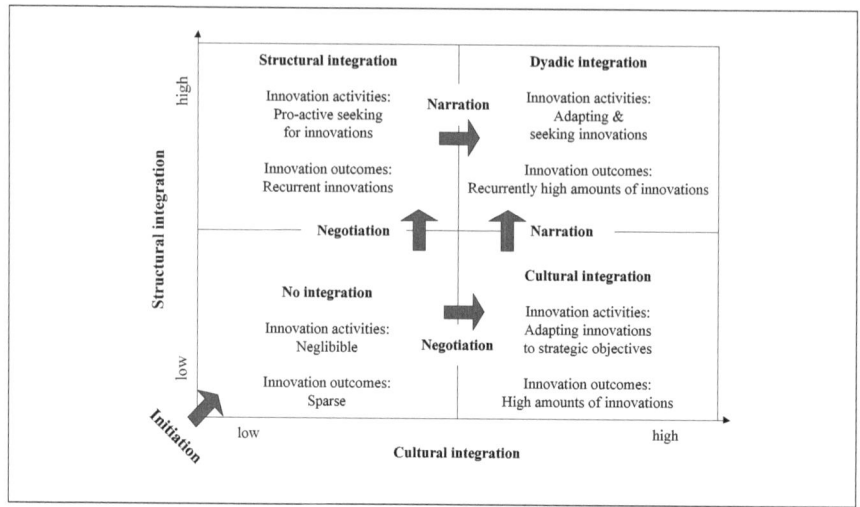

Figure 5: Four forms of organizational integration, based on transition strategies

Organizations attempting to anchor organizational innovation communities in their organizational structure may want to go beyond the still pre-dominant approach of a trial and error strategy with regard to the emergence and integration of such communities. To do so, findings suggest that an in-depth analysis of the current organizational integration of the organizational innovation community. By widening the predominant focus of how to (technically) set up organizational innovation communities to include the underlying aspects of organizational integration, these organizations will gain a deeper understanding of the applied innovation activities and the resulting innovation outcomes of organizational innovation communities. An analysis of organizational integration is therefore seen as a promising way for organizations to reduce the burden of implementing and using organizational innovation communities that do not generate the intended innovation outcome. Once organizations are aware of the de facto organizational integration of their organizational innovation communities they can benefit from the findings of what types of transition strategies exist and which transition strategy can be applied to move from one type of organizational integration to another. From the findings it is argued that this is a major challenge for organizations. Whereas decisions with regard to transition patterns so far have mainly been made by top management (Jarzabkowski, 2008), the research conducted proposes that transition strategies in the context of organizational innovation communities follow a 'bottom-up' process, i.e. a social movement starting from the bottom of the organization, climbing up the hierarchical realms and ending at top management conviction. However, it is argued that in addition to member-led decision making, managers may positively influence this bottom-up process by providing fertile seeding grounds for the increased anchoring of organizational innovation communities, such as that described in negotiation strategies. For instance, while managers may not be able to set up organizational innovation communities from scratch, they may have the ability to offer conditions in which important and unsolved issues ignite curiosity, reflecting important conditions for organizational innovation communities to occur.

Several limitations are recognized that could be addressed in future research. Given the nature of the study, the taxonomy developed should be seen as a structured analysis of reality, and not as the reality itself. First, whereas the findings rely on the analysis of organizational innovation communities in organizations which are 'innovation fore-runners', further research could increase the range of perspectives by studying organizational integration of organizational innovation communities and the related transition strategies in organizations which are at the beginning of their 'innovation journey'. Second, generalizations about if and if so how organizations can support the design of appropriate transition strategies are not made. Further research could examine whether the conscious design of transition strategies by the top management influences the pace at which organizational innovation communities are implemented and if such an approach would lead to (unwanted) hierarchical pressures within such communities. Thus, it is suggested that the influence of conscious top management interventions on distinct types of organizational integration would be a valuable topic of future research.

6 Conclusion

The main purpose of this part has been to investigate how the organizational integration of organizational innovation communities influences their innovation activities and the resulting outcomes. In addition, it aimed to understand what transition strategies exist to anchor organizational innovation communities within organizations. In other words, this part aimed to resolve Tom's struggle, as a top manager, in designing and implementing organizational contexts in support of organizational innovation communities.

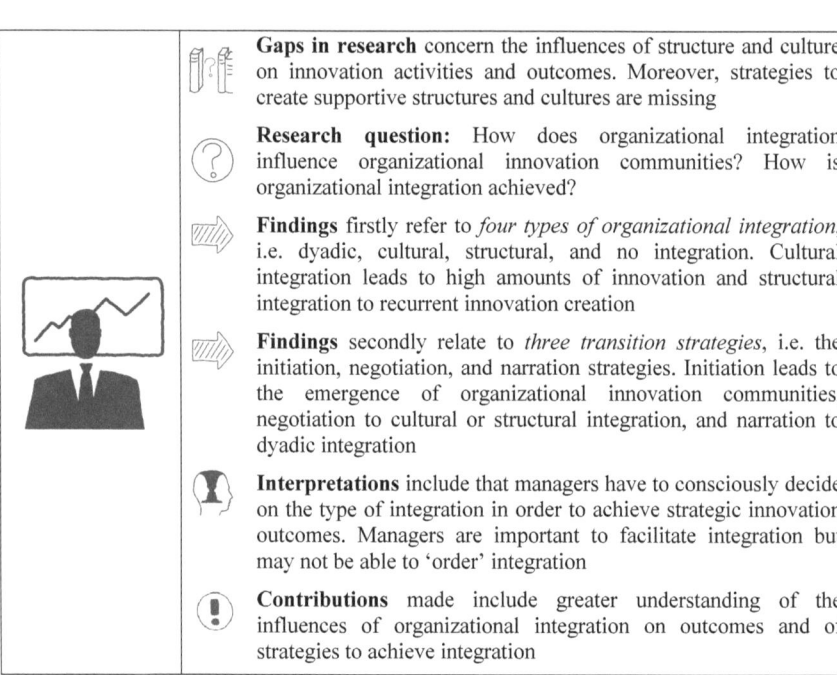

Gaps in research concern the influences of structure and culture on innovation activities and outcomes. Moreover, strategies to create supportive structures and cultures are missing

Research question: How does organizational integration influence organizational innovation communities? How is organizational integration achieved?

Findings firstly refer to *four types of organizational integration*, i.e. dyadic, cultural, structural, and no integration. Cultural integration leads to high amounts of innovation and structural integration to recurrent innovation creation

Findings secondly relate to *three transition strategies*, i.e. the initiation, negotiation, and narration strategies. Initiation leads to the emergence of organizational innovation communities, negotiation to cultural or structural integration, and narration to dyadic integration

Interpretations include that managers have to consciously decide on the type of integration in order to achieve strategic innovation outcomes. Managers are important to facilitate integration but may not be able to 'order' integration

Contributions made include greater understanding of the influences of organizational integration on outcomes and of strategies to achieve integration

Table 38: Taxonomy: Summary of part III

In sum, the twelve in-depth case studies enabled the development of a taxonomy of organizational innovation communities. This comprises four distinct forms of organizational integration, i.e., dyadic integration, cultural integration, structural integration, and no integration. Each form of integration is characterized by distinct sets of innovation activities and outcomes. Accordingly, the three unique transition strategies, i.e., initiating, negotiating, and narration strategy, contain clear guidelines to anchor organizational innovation communities culturally and/or structurally and have been integrated in the taxonomy. Seen in relation to an organization's strategy, which aims to anchor organizational innovation communities, the findings may be of major importance. It is believed that this in-depth qualitative work sets a foundation for future research that can extend insights about the underlying antecedents of the organizational integration of organizational innovation communities.

So far, the first major gap in research has been explored in this part, i.e., the influence of organizational contexts on innovation activities and outcomes in organizational innovation communities, as identified in the literature review. Since Tom now knows how organizational integration, by means of cultural and structural integration, influence innovation activities and outcomes and how organizational integration may be altered, the main focus is now directed towards Ina's major struggle, i.e., understanding how social processes unfold in the pursuit of innovation in organizational innovation communities. Hence, the next part provides a model to facilitate social processes for innovation development.

Part IV Sensemaking in organizational innovation communities (empirical study II)

1 Setting the stage[15]

	Literature suggests that weakly connected and cognitively distant community members collaborating have significant innovation potential
	Gaps in research concern understandings of how social processes for innovation development in organizational innovation communities unfold
	Insights concern understandings of social processes in organizational innovation communities for innovation pursuits

Table 39: Sensemaking: Existing literature, crucial gap, and insights

Now Ina's major struggle is put under the microscope. It deals with understanding social processes in the pursuit of innovation and ways to facilitate this process. She believes in the hypothesis that organizational innovation communities bear most potential when innovation development builds on joint meaning creation among participants with weak ties that are cognitively distant, as they gain access to expansive variety and divergent perspectives are integrated (Björk & Magnusson, 2009, p. 669; Bogenrieder & Nooteboom, 2004, p. 295; Nooteboom, 2000, p. 71; Brown et al., 2007; Ganley & Lampe, 2009). Despite the positive influence of these properties, they are constrained by lack of cohesion among participants, incomprehensibility of divergent perspectives, and hesitance to collaborate resulting from loose and irregular interactions, the occurrence of communication breakdowns, and resentment among community members (Leimeister et al., 2008, p. 353; Matei, 2004, p. 24; Bogenrieder & Nooteboom, 2004, p. 294; Björk & Magnusson, 2009, p. 669). Yet, there is still confusion about how to include expansive variety and divergent perspectives and to facilitate joint meaning creation in organizational innovation community settings effectively.

To gain insights on these issues, the question how innovation development actually unfolds collaboratively and which antecedents drive innovation development in community settings is analyzed. Although several antecedents are discussed in scholarly work that may influence innovation development, e.g., motivation (Hertel et al., 2003; Shah, 2006), boundary objects (Plaskoff, 2003; Fischer, 2001b), and knowledge processes (Bechky, 2003; Østerlund & Carlile, 2005), the main focus lies on the critical, yet understudied role of social processes for innovation development in organizational innovation community settings. To help facilitate this endeavor sensemaking theory is applied as an informative framework to analyze these social processes. In order to show the premises of this part, we will consider the following elements: 1) theoretical backgrounds, 2) methods applied, 3) empirical findings, and 4) a discussion of these findings in light of theoretical considerations.

---

[15] Part IV is based on a previous conference article discussed and presented (Bansemir, 2011b) at the 2011 & 2012 annual conference of the Academy of Management (AOM).

2 Theoretical perspectives

Sensemaking theory sheds light on collective attempts to construct meaning out of unexpected stimuli (Weick, 1995, p. 55; Starbuck & Milliken, 1988; Maitlis, 2005, p. 21; Balogun & Johnson, 2004, p. 524). Thereby, sensemaking provides a meaningful theoretical lens to study, analyze and understand social processes for innovation pursuits (Christiansen & Varnes, 2009; Hill & Levenhagen, 1995; Hill & Birkinshaw, 2010). Hence, Ina learns about how collaborative efforts in organizational innovation communities actually unfold in the pursuit of innovation. Understanding the joint meaning creation of weakly connected and cognitively distant community members might be particularly interesting for her. In the remainder of this chapter, major means and their influence on sensemaking in organizational innovation communities are theoretically discussed. In the following, major means of innovation development in organizational innovation communities are presented, before describing sensemaking theory in-depth. The table below summarizes and visualizes key theoretical insights.

 Sensemaking theory describes social processes through which meaning or understanding is collaboratively created in three steps, i.e. *stimulation, interpretation,* and *collaborative creation*

 A gap in research exists concerning how social processes in organizational innovation communities unfold and how they influence innovation development

 Key insights: Weakly connected community members [left] have the merit of information variety as a crucial brick for creativity and innovation to occur [right]. However, one major barrier related to weak connections stems from often lacking coherence [lower middle] among these community members

 Research question: How may weakly connected community members be *stimulated* to exert cohesion in organizational innovation communities?

 A gap in research concerns how social processes among cognitively distant community members unfold

 Key insights: Cognitively distant community members [left] bring-in divergent perspectives and hence possess significant innovation potential [right]. However, incomprehensibility and communication breakdowns [lower middle] still limit the effective inclusion of cognitively distant community members for innovation development

 Research question: How may cognitively distant community members bring-in their perspectives and *interpretations* for innovation pursuits effectively?

A gap in research relates to the nature of social processes through which joint meaning is actually created in organizational innovation communities

 Key insights: Joint creation of meaning, i.e. collaboration among several community members [left], is associated with the emergence of comprehensive understandings of complex issues and hence fuels purposive innovation development [right]. However, resistance of community members [lower middle] to collaborate often limit collaborative efforts

Research question: How does a community overcome resistance for the sake of *collaborative meaning creation* in organizational innovation communities?

Table 40: Theoretical background: Constraints of innovation development and sensemaking

2.1   Major means of innovation development

In the following several influences and constraints of innovation-related properties are described, namely 1) *weak ties*, 2) *cognitive distance*, and 3) *joint meaning creation*. First, the literature review (part II) shows that collaboration among participants with *weak ties* is particularly supportive for innovation to occur, but also that weakly connected participants are hampered by a lack of coherence. Weak ties are characterized by a low intensity of relationship between two or more individuals, compared to strong ties that are characterized by high relationship intensity (Granovetter, 1973, p. 1361). By means of fast information transmission from multiple sources bound in diverse social worlds, weak ties often result in the sharing of a greater variety of information (Amin & Roberts, 2008, p. 361; Lindkvist, 2005). Research frequently emphasizes the crucial role of information variety as a source of creativity and innovation (Amabile et al., 2005, p. 369; Fredrickson, 1998, p. 304; Isen et al., 1992, p. 66). Information variety delivers building blocks to find new combinations and hence is a major means for innovation to occur. For instance, literature on hybrid value creation emphasizes the importance of developing innovative solutions by combining services and products in an original way (Velamuri, 2011). As a consequence, weak ties deliver the variety of information necessary for innovation to occur (Assimakopoulos & Yan, 2006, p. 16; Björk & Magnusson, 2009, p. 669). However, because it is constrained by irregular, loose, and spontaneous interactions, collaboration among participants with weak ties is often hampered by a lack of cohesion (Bogenrieder & Nooteboom, 2004, p. 295; Björk & Magnusson, 2009, p. 669). Due to the low intensity of interactions, common objectives are unlikely to emerge. Overall, research points to the importance of facilitating interactions among weakly connected participants to capture both inherent innovation potential and effective collaboration, but this remains rather vague (see also part II).

Second, researchers agree that *cognitively distant* individuals in community settings bear the potential of significant innovations in terms of discontinuity and quantity, if the incomprehensibility of divergent cognitive frames is vanquished. Cognitive distance is characterized by little overlap of cognitive frames, for example resulting from dissimilarities of objectives, vocabulary used, professional education, and working styles (Nooteboom, 2000, p. 71; Boschma, 2005; Knoben & Oerlemans, 2006, p. 77). For instance, engineers and managers often have a different mindset, i.e. different cognitive frame. When looking at a machine, engineers might imagine the inner mechanisms that keep the machine working, while managers probably think of costs per piece, fixed costs and breakeven points. At the far end, cognitive proximity among individuals is characterized by similar knowledge stocks and experiences. By means of divergent cognitive frames, studies show that the collaboration of cognitively distant participants inherently exhibits notable innovative potential as perspectives are transferred from one domain to another (Amin & Roberts, 2008, p. 365; Hussler & Rondé, 2007, p. 300; MØrk et al., 2008, p. 12). For instance, transferring concepts in other contexts is a well-known strategy to innovate (Lakhani et al., 2007). Weakly connected community members are a particularly important means to effectively access concepts applied in other contexts. Problem-solving at Innocentive (an innovation intermediary) builds on this mechanism, as one key success factor is the attraction of solvers that create innovative solutions based on their knowledge in other domains of specialization. However, confined by incomprehensibility, communication breakdowns often inhibit expatiating divergent perspectives at an early stage of innovation development (Nooteboom, 2000, p. 71). Overall, research directs attention to the important, yet understudied, question of how to include divergent cognitive frames before communication collapses as a result of the incomprehensibility of different worldviews.

Third, studies show that *joint meaning creation* enables participants to strengthen innovation development, but often falls behind expectations because of the hesitance of participants. Joint meaning creation is defined by collaborative efforts of participants to develop comprehensive understandings concerning a given issue (Østerlund & Carlile, 2005, p. 101; Soekijad et al., 2004, p. 10; Palincsar et al., 1998, p. 10). Brown & Duguid's (1991) reflections on Orr's example of the copier machine workers shows that it is only by joint meaning creation that technicians, engineers, and maintenance workers are able to find out why the machine failed. As soon as the reason is found, solving the problem was only a minor step (Orr, 1986). By interacting with a wide range of information and a variety of divergent perspectives, an integrated and complete picture emerges that prioritizes the most urgent facets of any given issue (Brown & Duguid, 2001, p. 206; Soekijad et al., 2004, p. 10; Hemetsberger & Reinhardt, 2006, p. 210). However, joint meaning creation often remains limited due to resistance among participants (Schoberth et al., 2006, p. 250; Jones & Rafaeli, 2000, p. 1017). For instance, Bechky (2003) shows that joint meaning creation among engineers and manufacturing workers is often conflict-ridden and also leads to communication breakdowns. Conclusively, scholars hint at the importance of overcoming hesitance among participants as a major building block for joint meaning creation to emerge.

In line with these arguments, the positive effects of weak ties, cognitive distance, and joint meaning creation on the development of innovations, are accompanied with missing coherence, incomprehensibility, and a lack of joint meaning creation. To address these tensions, the main focus is on the critical – and surprisingly understudied – role of sensemaking for innovation development in organizational innovation communities.

2.2 Sensemaking as an interpretive framework

Sensemaking theory's fundamental premise includes interactive and, hence, social processes through which several individuals collectively attempt to construct and understand unexpected stimuli (Weick, 1995, p. 55; Starbuck & Milliken, 1988; Maitlis, 2005, p. 21; Balogun & Johnson, 2004, p. 524). Following sensemaking theory, three elements are of major importance relevance: 1) *stimulation,* 2) *interpretation*, and 3) *creation*.

First, *stimuli* ignite interactions among multiple individuals (Weick, 1995; Weick et al., 2005). They are defined as uncommon or unexpected events which are perceived by an individual as being significant and which ask for a response (Gioia & Chittipeddi, 1991; Gioia et al., 1994; Balogun & Johnson, 2004; Maitlis, 2005, p. 25). Researchers widely agree that only if stimuli include these characteristics, will individuals actively engage in sensemaking activities (Weick, 1995). For instance, Maitlis and Lawrence (2007, p. 78) show that if stakeholders perceived leaders as incapable of solving an important issue, they participated actively in sensemaking. Hence, stimuli are an important means of sensemaking, as they determine if, and if so to what degree, individuals engage in sensemaking activities.

Second, individuals *interpret* stimuli based on their cognitive frames, or in other words based on prior experiences (Weick, 1995; Maitlis, 2005, p. 25; Weick et al., 2005). Interpretation includes explicating observed stimuli and tagging them with clear labels (Weick et al., 2005, p. 411). Explicating is bound to individuals' cognitive representations, such as prior experiences, understandings, and vocabulary, and ignores differences among individuals (Weick, 1995, p. 110; Weick et al., 2005, p. 411): "Perceptual frameworks categorize data, assign likelihoods to data, hide data, and fill in missing data" (Starbuck & Milliken, 1988, p. 51). Weick is able to show the way firefighters make sense based on their cognitive frames, in this case with terrible consequences (Weick, 1993). In this example, a group of firefighters sensed that they would be in security as soon as they approached a nearby river. However they did not see that the fire was already crossing the river. As communication with the other firefighters was difficult, they had to rely on their own interpretation which led to 13 dead firefighters. In contrast, innovation researchers are able to demonstrate that variance in interpretations is a major building block for discontinuous innovations (Nooteboom, 2000, p. 71; Amin & Roberts, 2008, p. 365; Hussler & Rondé, 2007, p. 300; MØrk et al., 2008, p. 12). Altogether, interpretation is an individual cognitive process in which occurring stimuli are related to individual cognitive frames and explicated in terms of clear labels.

Third, *creation* refers to development of joint or collective meaning based on interactions among individuals and their interpretations (Maitlis, 2005, p. 21; Weick et al., 2005, p. 413). Researchers show that individual interpretations are negotiated, building on written or spoken communication channels, such as reports, graphs, schemes, conversations, presentations, etc. (Balogun & Johnson, 2004, p. 524; Gioia & Chittipeddi, 1991). Using these communication channels, individual interpretations are interactively discussed, rearranged and transformed in recurrent cycles until a sufficient level of collective meaning, adequately capturing all relevant perspectives, is created (Weick, 1995; Maitlis, 2005, p. 21; Bartunek et al., 2006, p. 183). These collective meanings are the main pillar for individuals to exert concerted activities, or in other terms, to act collectively (Maitlis, 2005, p. 21; Starbuck & Milliken, 1988; Weick, 1995). In Weick's example of the firefighters, the creation of joint meaning is considerably disturbed (Weick, 1993): One fireman sparks off a rescue fire and asks his subordinates to join him. However, as the other firefighters do not accept the superior role and still believe that they have to fight a small fire, a joint meaning is not created. Consequently only three of the firefighters who joined the rescue survived. In sum, collective meaning is created through intensive negotiation of individuals' interpretations and consistently leads to collective actions. Figure 1 visualizes and summarizes contemporary understandings of sensemaking and constraints in organizational innovation communities. However, Ina wants to gain more information about how to actually use the four steps to create innovations within organizational innovation communities. In other words, she wants to fill in the blank areas still visible in figure 6.

The link between stimuli, individual cognition, and collective meaning creation makes sensemaking theory a valuable interpretive framework for understanding innovation development in organizational innovation communities. Sensemaking theory is particularly applicable to explain the above-mentioned issues of missing coherence, incomprehensibility, and the lack of joint meaning creation, as it offers insights on how the social processes of innovation development actually unfold and how these challenges are tackled. It clearly underlines this issue, emphasizing that the interplay of the individual's interpretation and action take place in social contexts, are triggered by uncommon or unexpected stimuli, and are built on collective meaning creation. For instance, differences in stimuli may explain why community members with weak ties fail in executing tasks efficiently. Using sensemaking theory as a perspective to analyze how innovation development in organizational innovation communities unfolds, a more nuanced understanding of antecedents for successful innovation development is derived. Specifically, a process that better explains the interrelations between social dynamics in innovation development and innovation outcomes is proposed.

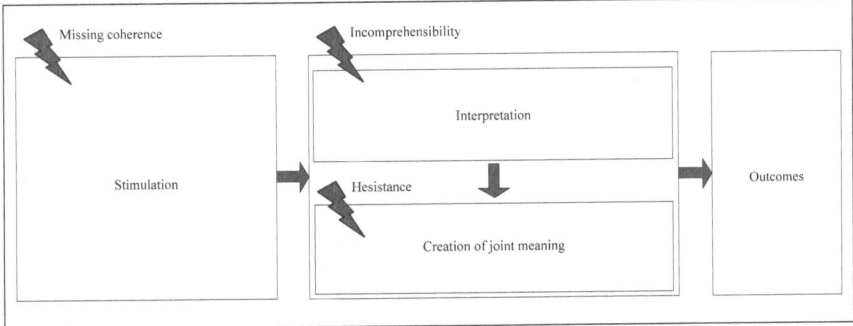

Figure 6: The social process of sensemaking

3 Research methods

A sensemaking perspective offers significant opportunities to comprehensively understand actual collaborative innovation development in organizational innovation communities. However, one challenge with using sensemaking theory in an empirical setting stems from the fact that sensemaking is a procedural theory (Maitlis, 2005), including barely observable individual cognitions (Weick, 1995). These characteristics of sensemaking as a theoretical perspective make design science a particularly meaningful underlying research paradigm as the IT artifact may be more accurate in mapping these barely observable individual cognitions. The following sections describe the applied action research method as a vehicle to investigate sensemaking for innovation pursuits in organizational innovation communities.

3.1 Research design

Action research as a method is especially suitable to derive models concerning the efficient use of innovation IT artifacts as described in design science (Hevner et al., 2004, p. 82). Specifically, action research methods conform to the iterative generation and test cycles stipulated by design science. Action research applies recurrent cycles of *generating and refining models* (Eden & Huxham, 1996) in field settings and in the pursuit of constraint resolution (Lüscher & Lewis, 2008, p. 221). Practically, in the present, constraints for innovation development in organizational innovation communities are cyclically identified, explanations derived, and solutions applied. Consequently, iterative cycles of generating and refining models are a meaningful mode of generating an in-depth understanding of sensemaking in organizational innovation communities.

In other words action research may also be seen as a series of qualitative and explorative field experiments, i.e. experiments in 'real-life' organizational settings, tackling actual constraints. In this logic, constraints identified from actual use of the IT artifact give a basic impetus to develop more nuanced models. For instance, at one point in time the engagement of participants in the pilot organizations almost completely vanished. Understanding this behavior and finding models to fuel momentum again has been critical. Moreover, applying these models and assessing their effectiveness recurrently was a meaningful research strategy to test and refine these models until saturation was achieved, i.e. until the new knowledge gained by conducting another cycle of exploration was minimal. Hence, action research as a series of qualitative and explorative field experiments delivers results that may exert high validity and generalizability (Christensen, 2007; Witte, 1972).

According to experimental methodology, the conducted action research was carefully planned in terms of the selection of participants, tasks, and overall organization (Christensen, 2007; Stanley & Campbell, 1966; Maxwell & Delaney, 2004). First, compared to most experimental studies, the sampling of participants was not randomly conducted but consciously designed. As the main aim is to draw conclusions about social processes among weakly connected and cognitively distant employees for innovation pursuits, a criterion for inclusion was that participants in the Open-I pilots should have little overlap in associated departments and professional backgrounds. Studying these employees was especially interesting as they had the merit of innovation but also faced severe constraints (see literature review). Second, literature emphasizes that joint meaning creation in organizational communities leads to superior results (Østerlund & Carlile, 2005, p. 101; Soekijad et al., 2004, p. 10; Palincsar et al., 1998, p. 10). To address this aspect, pilot organizations were explicitly asked to identify innovation tasks that have been unanswered for several years. In other words, pilot organizations had been unable to solve the main innovation task given to participating employees within hierarchical realms. Third, constraints identified from the pilots and derived models to face them were a major trigger for the series of experiments. For instance, at an early stage of the pilots constraints of missing coherence, resulting in a lack of engagement, emerged. Finding solutions for this constraint has led to a series of experiments in which the effectiveness of the derived models is tested. Consequently, the carefully planned pilots were a meaningful mode to address the main research question. The subsequent section describes these crucial design parameters of the applied action research in greater detail.

3.2   Sample and data collection

The **sample** consists of participants, acquired in three independent empirical settings, in large service firms in Germany introducing organizational innovation communities by means of the developed Open-I platform. Within these organizations in-depth longitudinal exploration of organizational innovation communities starting from 2007 was conducted.

Besides early preparation activities, the *first pilot* has been in operation since March 2009. Conscious selection according to the above-mentioned criteria, led to the inclusion of 37 employees from heterogeneous professional backgrounds, different departments, and varying hierarchical levels. All of them may be described as 'ordinary employees' in terms of their prior experiences concerning innovation development. Rewards (financial or other) were not promised at the beginning or given at the end of the pilot, nor were outcomes of the project related to goal achievement talks to ensure voluntary participation. During piloting, the organization underwent a dramatic change, sparked by a major revision of the organization's main service. As a result, the project experienced low management prioritization, in terms of time and budget given, but it was under intensive pressure to deliver valuable results. The *second pilot* started operations in February 2010, including 31 employees in a similar composition to the first pilot. It is important to mention that the financial crises of 2009 hit this organization considerably: one fifth of all employees lost their jobs. Hence, the pilot experienced a challenging environment in terms of prioritization and accessible resources. The *third pilot* was initiated in January 2011, including 19 employees from different backgrounds, geographic locations, and hierarchical levels. Altogether, this study builds on intensive experiences and cyclical experimentation within three independent large organizations including 87 employees.

The **procedure** of innovation development takes place in workshops on the organizations' sites as well as on the developed Open-I platform. Onsite workshops are conducted at the beginning, midway, and at the end of one development cycle, flanking community collaboration. However, most innovation development activities are related to the work on the organizational innovation community as a main focal point. The Open-I platform provides common community functionalities, e.g., individual profiles, interest groups, etc., and innovation support functionalities, e.g., innovation profiles, tech-clouds, etc. Additionally, the platform supports collaborative innovation generation, refinement of innovation and peer-evaluation. Moreover, one design element of the platform which fostered innovation development was the innovation points automatically calculated by and displayed on the platform. They were given for activities, concerning exchange of opinions, sharing ideas, and commenting on concepts. As a result, rich qualitative data is triangulated with descriptive quantitative measures.

The main **task** was to create original and applicable innovations for an unsolved issue and was split in three subsequent steps, i.e., innovation generation, refinement, and evaluation. This process was adopted from the workflow procedures typically applied for structured innovation development in organizations. First, participants were asked to generate a considerable number of rough and unstructured ideas. This process was initiated within a first workshop, in which main functionalities of the community platform supporting this stage were also explained. Second, while merging and refining these ideas, more sophisticated concepts were developed. Functionalities supporting this step were introduced and explained in the second workshop. Third, participants evaluated concepts submitted by other participants, using various evaluation methods. Results from this stage were presented during a public open space meeting. Constraints in all three steps were a major means to understanding the complete cycle of innovation development in organizational innovation communities.

To develop in-depth understanding of action research interventions and to ensure rigor and relevance, **data collection** was longitudinal, triangulated, and intensive. First, long-lasting and trustworthy relationships existed with employees of the studied organizations, on various hierarchical levels, reaching from top management to employees with no leadership responsibility. Additionally, all organizations were closely accompanied, resulting in data that captures all significant events related to these communities from the outset.

Second, the triangulation of data included data collection from varying sources and applying different methods. This includes *observations, formal* and *informal interviews,* and *descriptive quantified measures. Observations* of actual innovation development during collaborative activities on the community platform formed the main basis for data analysis and served as a focal point to which other data relate. They refer to virtual workshops, using the community platform for actual idea generation and innovation development. A virtual whiteboard built into the community platform constituted the main functionality for these workshops. All participants saw the same virtual whiteboard with messages, ideas, etc., from other participants popping up in real time. Additionally, they were able to edit others' ideas, join multiple ideas into one, and group related ideas into clusters. During these workshops (lasting for one hour), participants were located at their regular workplaces and were connected using a telephone conference solution. Altogether 28 virtual workshops were conducted. In each workshop, extensive notes about the course of actions were taken, outcomes recorded, and documents gathered. Moreover, 22 virtual workshops were videotaped: all activities of participants on the virtual whiteboard and all oral communication via the telephone conferences were caught. Additionally, after each virtual workshop, brief reflection meetings including responsible managers were conducted. Moreover, discussion rounds on evolving issues were accomplished with participants, responsible managers, researchers, and top managers. Observations provided rich procedural data, particularly witnessing participants engaging in innovation-related sensemaking and observing circumstances supporting or hindering this process. The following figure shows the virtual whiteboard as a major means of observations.

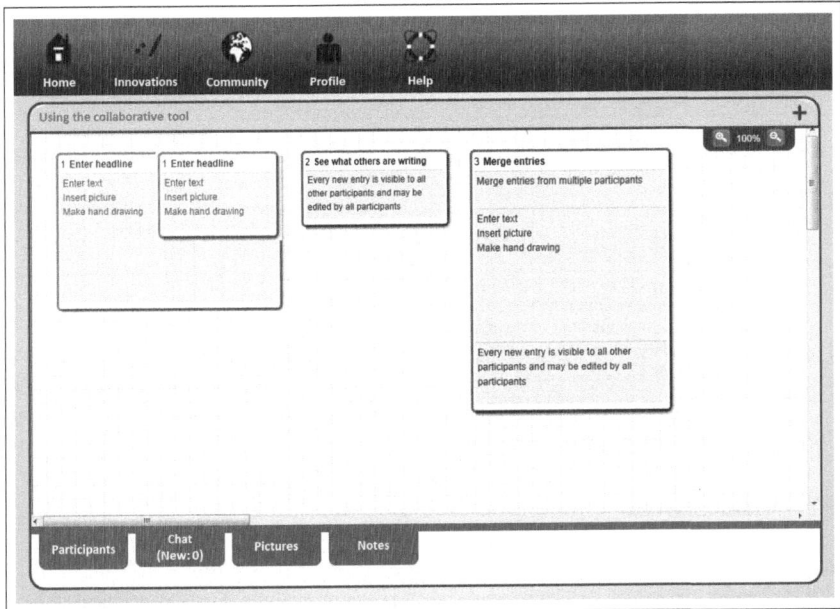

Figure 7: Screenshot of the virtual whiteboard used for virtual workshops and observations

Moreover, in each organizational setting, *formal interviews* before and after the first cycle of innovation development were conducted with responsible managers but also with participants and top managers. Additionally 24 semi-structured interviews, of which all were recorded and transcribed, were completed. Moreover, many (around 150) *informal interviews* were conducted throughout innovation development – with participants, managers in charge, and top managers – as opportunities arise. To capture important insights notes were taken immediately after interactions or, if possible, during the conversation itself. These informal interviews captured spontaneous and straightforward opinions about recent events or the project at large. As a result, access to experiences and opinions which are usually unamenable to investigation were gained. Finally, supplementing these rich data, the community platform provided additional *descriptive quantified measures*. For instance, the number of ideas created or innovation concepts developed were automatically calculated. Furthermore, participants collected activity-related innovation points. As a result, large amounts of rich qualitative data triangulated by descriptive quantitative measures comprehensively capture significant events from multiple sources.

Third, miscellaneous other data sources from *intensive relationships* were used to capture data. For instance, intensive exchange via email (more than 542 topic-related emails were documented) and weekly telephone meetings were established to discuss pressing issues and recent developments mainly, but not exclusively, with responsible managers. Moreover, data collection spanned numerous professional short-term involvements, such as conducting topic-specific presentations, participating in meetings, and leisure activities, etc. Consequently, trustworthy relationships and honest and meaningful exchanges that went beyond professional duties were established. The following table summarizes data collection.

Type of data	Description	Amount
Observation	Observation of actual innovation development during virtual workshops using the virtual whiteboard, built-in the Open-I platform.	22 videotaped observations lasting for 1 hour
Formal interviews	Formal interviews using a semi-structured interview guideline with top managers, responsible managers, and participants before and after piloting	24 recorded and transcribed interviews
Informal interviews	Informal interviews with top managers, responsible managers, and participants mostly during piloting as opportunities arose, for instance at joint lunches	~ 150 informal interviews
Quantitative measure	Platform-provided descriptive quantitative measures concerning number of innovations created and activities of communities on the platform	2 types of measures in 3 organizations
Intensive relationship	Intensive relationships indicated by the number of emails, but also leisure activities, such as joint dinners	542 emails and 12 private meetings

Table 41: Overview of data sources

3.3   Data analysis

In general, data analysis entails interrelated stages on two distinct levels to meet practical as well as research standards: 1) immediate and 2) posterior analysis. First, as practical issues emerged which were unplanned and most often needed quick responses, an immediate, interactive and spontaneous approach was applied. This approach included detailed and prompt discussions and reflections among researchers and practitioners. Typically, a team of six researchers and practitioners was involved, but it was also extended to the whole core team of researchers and practitioners if necessary. These rather spontaneous reflections and discussions of emerging issues proved to be an informative way to gain in-depth knowledge about major influences for the successful design of sensemaking processes and possible solutions if obstacles emerged. This approach relates to Torbert's (1976) call that action research should be "useful to the practitioner at the moment of action rather than a reflective science about action" (Reason, 1988). Second, to base model development on more solid grounds and to gain a nuanced understanding, posterior analysis included rigorous application and documentation of data analysis procedures, as propagated by Eden & Huxham (1996, pp. 78–79). While both modes of data analysis follow similar steps and results from either level mutually fueled model development, procedures of the posterior analysis are displayed in the following.

Posterior data analysis included the following stages: preparing data, identifying issues, and developing theory. First, *preparation of data* included organizing all data, in terms of data source, methods used, appearance in the timeline, etc., and transcribing all taped data material. To ensure that transcripts captured all important information, expressions such as emotional states were transcribed where necessary. Transcripts from the video data alone exceed 300 pages of written documentation. Second, while coding data using Atlas.ti and reflecting this information with experiences within the organizations, relevant text passages and *identified issues* of significant relevance are highlighted. This stage was conducted by three independent coders to check for possible threats of objectivity and for inter-coder reliability. Labels and anchor examples were developed for each new issue, to attain a common understanding of coders. Third, *theory development* referred to bundling, summarizing, and explicating identified issues in a process of defamiliarization, in the sense that rich data material was revised in recurrent cycles and reduced to the most relevant contents (Eden & Huxham, 1996, pp. 78–79). Hence, the labels as well as contents of developed models changed as more data was included. This procedure is partly informed, but not restricted, by factors identified in community literature (as presented in the literature analysis), in the sense that they are included and linked to the interpretations. However, due to space constraints the main focus refers to issues rarely described previously. To profit from action research approach most, dynamics of interventions are taken into account.

4 Findings

The research question refers to how sensemaking actually unfolds in the pursuit of innovation development in organizational innovation communities. Applying an action research approach in four empirical settings, analysis discloses a distinct set of factors. The following sections describe findings related to 1) coherence, 2) cognition, 3) collaboration, and 4) outcomes. Building on these results a model of sensemaking in organizational innovation communities is developed in the discussion section.

Frist-order concepts	Second-order construct	Aggregated dimensions
Collaborative challenges	Mastering challenging situations	Coherence
Content challenges		
Specifying a concrete issue	Igniting purpose	
Maintaining a broad focus		
Perception of priority	Perception of usefulness	
Perception of effectiveness		
Associative writing	Stimulated creativity	Cognition
Subliminal stimulation		
Percipience of others' contributions	Subliminal competition	
Pro-active encouragements		
Jokes	Perpetuated activities	
Flow		
Discovering interrelations	Interactive transformation	Collaboration
Determining clusters		
Identifying contradiction	Contradiction resolution	
Mutual resolution of contradiction		
Intractability of contributions	Pragmatic transformation	
Reversibility of clustering		
Perception of transparency		
Project initiation	Competitiveness	Outcomes
Direct comparison		
High numbers of ideas	Number of innovation ideas	
Discontinuity of ideas	Scope of innovation ideas	
Overall motivation	Latent effects	
Indirect application		

Table 42: Summary of aggregated dimensions, second-order constructs, and first-order concepts

4.1 Coherence

Data analysis reveals that establishing coherence among multiple community members creates stimulation as described in sensemaking theory. Three main notions were found to determine coherence among individuals to endure their ambitions in innovation development: 1) a challenging situation, 2) igniting purpose, and 3) usefulness of purpose.

	Coherence characterizes situations in which multiple, also weakly connected, community members share objectives
	Findings suggest that coherence among weakly connected community members [right] is likely to emerge if community members jointly master *challenging situations* [upper left], have a purpose which is *igniting*, i.e. community members are able to connect to this purpose emotionally [middle left], and perceive themselves as useful contributors [lower left] 
	Ina **learns** about major influences on the emergence of coherence among weakly connected community members
	Project example: Introducing virtual and collaborative workshops proved to be crucial to achieve coherence among weakly connected community members. After several rounds of trial and error, the first such workshop that worked well concerned an innovation challenge identified as being crucial: to develop a service atmosphere which has 'flair'. The following introduction was given: The example of Starbucks was identified to capture all relevant aspects of the concept 'flair': A lot of people go to Starbucks not simply because of the quality of the coffee but because they feel like they are 'at home' in their living room. This 'flair' captured the essence of what the virtual creativity session was about quite well. Consequently, the introduction sounded something like this: 'You see the picture of the coffee cup, right? [a picture of a Starbucks coffee cup was shown on the Open-I platform] Who of you go to Starbucks because of the quality of the coffee? [Some participants say something like 'yes sure, but actually I also feel at home there.'] So, a lot of people go there because it feels homey, it feels like sitting in your living room. In other words, they go because of the flair or atmosphere. This is actually an example of what we were aiming for in this session: We want to think about crucial elements to improve the flair or atmosphere in our waiting rooms so that customers feel better.' By giving such a rich example every participant had a similar and appealing picture of the main objective of the session in mind *(igniting purpose)*. The objective was indeed challenging *(i.e. it was a challenging situation)*, but because every participant had been to Starbucks at least once they also felt they could be useful contributors *(usefulness of purpose)*.

Table 43: Coherence: Definition, findings, learning, and project example

First, the analysis revealed that sensemaking within the organizational innovation community is supported by mastering *challenging situations*. This is a second-order construct that includes the following first-order concepts: 'collaborative challenges' and 'content challenges'. Table 1 displays exemplary quotations from data, illustrating each first-order concept. For instance, quotation 1.1.1 shows that evolving technical issues challenge participants. While faced with these issues and resolving them, participants develop solidarity: One participant had an issue related to resizing a text box. As the quotation shows, all other participants responded immediately and jointly helped. In the following other participants explicated issues in a similar vein and were provided with helpful comments from fellow participants.

Second, *igniting purpose* relates to the conditions that enable participants to connect to a given issue and to openly explicate connotations. This second-order construct includes two major first-order concepts: 'Specific issue framing' and 'broad focus'. First, quotation 1.2.3 illustrates the positive influence of specifically framed issues: Before starting a virtual session, several pictures for illustration were shown which led to a precise understanding of the main issue. In contrast, abstract issues were particularly challenging for participants and led to the abortion of activities (quote 1.2.1). Besides addressing a specified issue, quotation 1.2.2 also shows that a broad focus enables participants to bring in a broad spectrum of individual connotations associated with a given issue.

Third, *perception of usefulness* refers to individuals' feelings that engaging in innovation development is reasonable. Two first-order concepts characterize perception of usefulness: 'Perception of priority' and 'perception of effectiveness'. In quotation 1.3.2 and 1.3.3, participants explain that participation remained drastically below expectations as the project obtained low priority from team leaders: 'I felt very motivated, but my superior asked me to do so many other things' (quote 1.3.1). To increase perception of priority, participants proposed that the moderator should give homework. Quotation 1.3.2 exemplarily displays the concept of 'perception of effectiveness': Participants exerted significant problems with executing an innovation development cycle because they did not perceive themselves to be valuable contributors.

No.	Quotations concerning coherence
Collaboration challenges	
1.1.1	P3: 'For me, it does not work.' P1: 'For me, it does.'[All participants laugh in conjunction] P4: 'It is like Power Point.' P1: 'Go to the lower right corner.' P2: 'A double arrow appears.' P3: 'I see, wonderful!'
1.1.2	P1: 'Oops, could you put all boxes to the left, 'cause I'm sitting at my netbook and everything is way too small, I do not see anything.' P2: 'We will close the big one, we do not need it anymore.' P1: 'OK. Can you do it [P2]? [...] Now it is on top!' [...] P2: 'On top? Yes, I mean, you can also use the scroll bar to get there! No? You know the scroll bar, right?' P3: 'Yes, yes, sure. [...] Yes, it is good this way.'
Igniting purpose	
1.2.1	P2: 'I also have the problem that I do not really understand the topic, but anyway [...]' P3: 'You do not understand the topic?' P1: 'The headquarter, what is the headquarter?' P3 and P4: 'We are!' P2: 'We? ... as we provide the information?' P3: 'Exactly, we casually termed those who provide information 'headquarter'.' P2 and P1: 'OK, let's see ...' (R)
1.2.2	P4: 'Oh yes, I see 'bicycle' and so ... ah, now there also appears 'bicycle repair'. Excellent!' P6: 'I like them as well.' P5: 'Yes, I really like them too. Unbelievable. 'Travel provisions'! Yeah.'
1.2.3	P1: 'Today, we are doing a short [virtual] city tour. I will guide you through the city. I really like doing this [...]. But today I will explain to you the major sights, we just look to the left and to the right of the street. OK. If we take a look ... You may also click on the pictures to enlarge them. You can also leave a comment. Everything that comes to your mind. ... We leave our main building in inhospitable weather and just look around this area and the first thing we see is an advertising pillar 'Mut zur Farbe'. [pictures of display windows at the city center are displayed] We are leaving our city tour and turn to the whiteboard [...]' P2: '[...] Now, we are searching for the similarities between these businesses, in other words how to cluster the businesses we have seen on our city tour.'
Perception of usefulness	
1.3.1	I5: 'I felt very motivated, but my superior asked me to do so many other things'
1.3.2	I2: 'What I experienced as useful was when we had to do homework. If someone gave the advice, now you have to do something. Then I found it useful to contribute'
1.3.3	I3: 'I have to admit, what I found useful was that you were able to log in, click around and check 'can I contribute to a task?' However it is necessary that there is a task that clearly shows that there is something at stake.'

Table 44: Quotations concerning coherence

4.2   Cognition

Two major means are found to determine the cognition of individuals while traversing an innovation development cycle: 1) stimulated creativity, 2) unconscious competition, and 3) perpetuated activities.

	Cognition refers to the expression of community members' interpretations and associations while confronted with a given innovation challenge
	Findings indicate that cognition is most effective for innovation development [right] if creativity is stimulated, e.g. by means of typing what comes to the mind (associative writing) [upper left], if community members compete unconsciously [middle left], and if creative cognitions are perpetuated, e.g. in terms of jokes or flow [lower left] 
	Ina **learns** that capturing the individual cognitions of cognitively distant community members is facilitated by three major influences, i.e. stimulated creativity, unconscious competition, and perpetuated activities
	Project example: The screenshot below displays individual cognitions in the form of text written on the virtual whiteboard during a virtual workshop session. It may be interesting for the reader how these results were achieved in a one hour virtual meeting: After the introduction of the moderator, community members started to write down their thoughts, often for more than ten minutes. Even though a telephone conference was established, the only thing you could hear was the clattering of keyboards (stimulated creativity). Hearing other community members working and seeing others' text boxes popping up in high frequency led to an atmosphere in which all participants somehow competed concerning the number of text boxes (unconscious competition). After approximately ten minutes, creativity often reduced, which was obvious because fewer text boxes were submitted. However, a joke at that point often led to loud laughter and afterwards to an increase of text entries again.

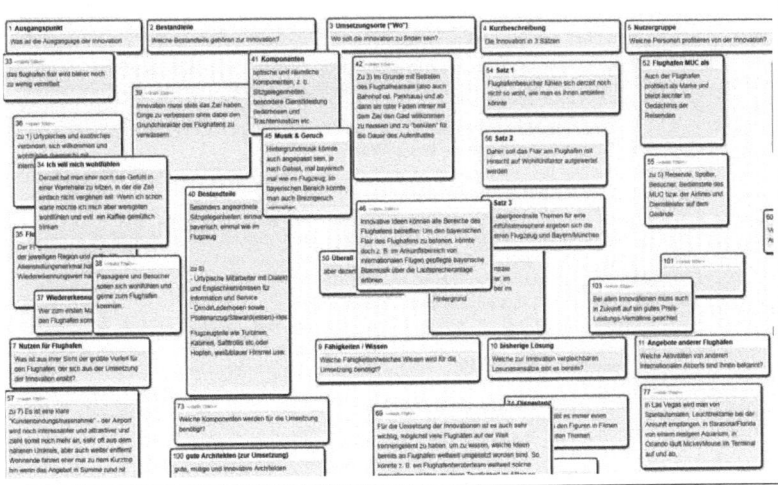

Table 45: Cognition: Definition, findings, learning, and project example

First, *stimulated creativity* refers to conditions that enable participants to freely and fluidly explicate large quantities of evolving connotations concerning issues at hand. This second-order construct includes the two first-order concepts 'associative writing' and 'subliminal stimulation'. In quotation 2.1.2, all participants started to write down their connotations related to the given issue, whether it made sense or not, which generated a decent amount of – as yet, fuzzy – ideas. Additionally, quotations 2.1.1, 2.1.3, and 2.1.4 highlight the positive impact of subliminally seeing others' ideas stimulating the natural flow of ideas. It shows that participants are cognitively stimulated by others' ideas even though these ideas remain at a subliminal level.

Second, *unconscious competition* describes conditions in which participants are passively or actively triggered to exert high levels of activities. Unconscious competition is determined by two first-order concepts: 'percipience of others' contributions' and 'pro-active encouragements'. Quotation 2.2.1 demonstrates situations in which participants increase their individual efforts as a consequence of passively noticing others' contributions, whether auditory or visual. Hearing clattering noise from others' keyboards and seeing others' contributions popping up in real-time significantly affected efforts in a positive way. Moreover, pro-active encouragements relate to direct feedback concerning specific ideas or the overall level of activity and have a positive impact on individuals' levels of activity. Quotations 2.2.3 and 2.2.2 elucidate these two aspects.

Third, *perpetuated activities* concern conditions that elongate high levels of activity. Perpetuating activity is positively influenced by the following first-order concepts: 'jokes' and 'flow'. An excerpt of the virtual workshop 4 as presented in quotations 2.3.1 and 2.3.2 illustrate the positive impact of jokes for activities in circumstances when concentration is about to flatten out. In this kind of situation, jokes seem to provide the needed variation to elongate cognitive processes. Moreover, keeping interruptions or disturbances away from participants – for example explaining technology midway through the creative process or introducing a completely new issue – allows them to stay in a fluid mode of activity. Quotation 2.3.3 shows that constant interruptions constrain cognitive processes. However, quotation 2.3.4 shows that sparse interruptions do not impede innovation development.

No.	Quotations concerning cognition
Stimulated creativity	
2.1.1	P4: 'Oh yes, I see 'bicycle' and so ... ah, now 'bicycle repair' also appears. Excellent!' P6: 'I like them as well.' P5: 'Yes, I really like them also. Unbelievable. 'Travel provisions'! Yeah.' **P4: 'Now, I am also getting ideas. Wait [...]'**
2.1.2	P1 [entering text box no. 6]: 'overwhelming many people in a small room with information' **[min. 17; sec. 32]** P2 [entering text box no. 8]: 'bad food, fewer **breaks**'[min. 17; sec. 34] P3 [entering text box no. 7]: 'fewer **breaks**, no discussion, no exchange of experiences' [min. 17; sec. 34] **P2** [opening text box no. 9; **min. 17; sec. 39**] **P1** [opening text box no. 10; **min. 17; sec. 39**] P4 [opening text box no. 9; **min. 17; sec. 44**] P3 [opening text box no. 9; **min. 17; sec. 45**] **P1 [entering text box 10]:** 'always the same people that already know everyone.' **[min. 18; sec. 02]**
2.1.3	P1: 'First, I have to read what you wrote' [smiling benignly; min. 19; sec. 29]
2.1.4	P3: 'But, that was also **telepathy**, you thought about the signs and I about the rooms.' [...] P2: '... and it fits!' [laughs] P1: 'Yes, it fits very well. Wonderful. So, it was not a software failure, but **telepathy**. Wonderful. I will get back to work.'

Unconscious competition	
2.2.1	P3 [enters heading of text box no. 9]: 'Service' [min. 14; sec. 17] **P2 [enters heading of text box no. 8]: 'Workplace' [min. 14; sec. 23]** P4 [opening text box no. 13; min. 14; sec. 29] **P2 [enters heading of text box no. 10]: 'Parking lot' [min. 14; sec. 30]** P1, P2, P3, P4, P5 and P6 [clattering noises from **heavy typing**] **P2 [opening text box no. 14; min. 14; sec. 33]** P3 [enters heading of text box no. 3]: 'First impression' [min. 14; sec. 34] P1, P2, P3, P4, P5 and P6 [clattering noises from **heavy typing** for the next couple of minutes] […] P4: '**It is really hard to keep up with you [P2]!'**
2.2.2	P1: 'You see, you do not need hundreds of participants; thoughts bubble up with fewer people anyway. **Wonderful!'**
2.2.3	P7: 'Honestly speaking, you really did an amazing job. I have never seen that the whiteboard so nearly full [no space available]. So far, you have also achieved more than 100 ideas in such a short time. Well done. I am really amazed and overwhelmed!'
Perpetuated activities	
2.3.1	P5: 'That's not bad! It's slip-sliding away' [chuckles] P5, P7 and P8: [laugh]
2.3.2	P1: '[…] I always start with the text and afterwards think about the heading' **P1, P2, P3, P4 and P5 [loud laughter; min. 19; sec. 02]** P2: 'I am doing it also that way …' **P1, P2, P3, P4 and P5 [loud laughter again; min. 19; sec. 06]** P3: 'But, that was also telepathy, you thought about the signs and I about the rooms.' P1, P2, P3, P4 and P5: [loud laughter again; min. 19; sec. 08] P2: '… and it fits!'**[laughs]** P1: 'Yes, it fit very well. Wonderful. So, it was not a software failure, but telepathy. Wonderful. **I will get back to work.'** P2: 'Yes.'**[still laughing]** P1 [enters heading of text box no. 25]: 'coffee'**[min. 19; sec. 28]** P4 [opening text box no. 26]: 'coffee' [min. 19; sec. 52] **P1, P2, P3, P4 and P5: [considerable activity for next 8 min.]**
2.3.3	P2: 'So, this is my first idea.' P1: 'Yes. I see it.' P2: 'Ah. Where? Ah, you have it on the screen. Ah, last time at the workshop the font size was problematic. Sometimes it was too small or too big. This time it is good. Is it also possible to move the text box? I cannot do this. Is it possible?' P1: 'Yes, keep the left finger on the left mouse button and move the box.' P2: 'So go to the text box. I hold the left finger on the left mouse button and now I move it. Ah, it works; I see the text box moving.'
2.3.4	P1 [entering text box no. 10]: '[…]' [min. 13; sec. 44] **P2: 'You can also add a heading …'** P1 [opening text box no. 12; min. 13; sec. 48] **P2: '… on top. Then you know more or less what it is about.'** P3 [entering text box no. 11]: '[…]' [min. 13; sec. 55] P3 [enters heading of text box no. 11]: '[…]' [min. 13; sec. 58] P1 [entering text box no. 12]: '[…]' [min. 13; sec. 58]

Table 46: Quotations concerning cognition

4.3   Collaboration

Data analysis reveals two key notions, positively influencing collaboration among participants while developing innovations: 1) interactive transformation, 2) contradiction's transformation, and 3) pragmatic transformation.

Collaboration describes interactive processes in which community members transform their individual meaning into collective meaning

 Findings indicate that collaboration is most effective for innovation development [right], if community members interactively transform individual interpretations, for instance while discovering interrelations and building clusters [upper left], if community members resolve contradictions, including identification and resolution of contradictions [middle left], and if this transformation is done pragmatically [lower left]

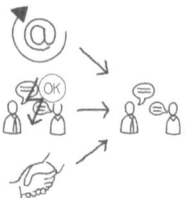

 Ina **learns** how joint meaning creation builds on interactive transformation, contradiction resolution, and pragmatic transformation

 Project example: At the end of each virtual workshop individual interpretations (developed in the stage 'cognition') were connected and clustered to sort the huge number of first ideas. The result of such collaborative efforts is represented in the pictures below. On the left side all ideas are shown as entered on the platform. On the right side ideas are connected and result in three major clusters.

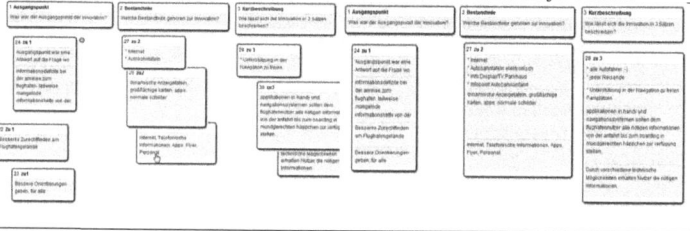

Table 47:Collaboration: Definition, findings, learning, and project example

First, *interactive transformation* concerns collaborative processes in which participants' labels, or in other words written associations, are collectively converted into sound concepts. Labels are combined while traversing a cycle of two steps, including 'discovering interrelations' and 'determining clusters'. In quotations 3.1.1 and 3.1.2 it becomes clear that participants try to find interrelations between various labels written on the whiteboard. Subsequently, quotations 3.1.3, 3.1.4, and 3.1.5 describe the manner in which various associations are combined and visually organized, following content affiliations.

Second, *contradiction transformation* designates processes in which participants jointly overcome paradigms, i.e. widely accepted and applied ways of thinking, opposing new insights from community activities. This second-order construct is characterized by two major first-order concepts: 'identifying contradictions' and 'mutual resolution of contradiction'. Foremost, quotations 3.2.1 and 3.2.2 describe an interactive discovery of obsolete paradigms. Besides the fact that participants figured out that some ideas already existed, they had not been addressed effectively in the organization. Hence, new ways of thinking were required, i.e. they identified an obsolete paradigm. Moreover, quotations 3.2.1 (last sentence) and 3.2.2 show that immediately after identifying the obsolete paradigm, participants start to search for reasons and thereby find possible resolutions for the initial contradictory situation.

Third, *pragmatic transformation* describes processes in which participants pragmatically transform interpretations, i.e., aiming for the best solutions ignoring potential political bias or personal offense. This second-order construct is characterized by three major first-order concepts: 'lacking traceability of contribution', 'reversibility of clustering', and 'perception of transparency'. In records 3.3.1 and 3.3.2 it becomes clear that contributions were not traceable to one particular participant. This fact led to participants mixing and matching contributions along content affiliations while disregarding the source of contributions. Also, quotation 3.3.3 shows that the possibility to reverse clusters supported a pragmatic approach, even though this functionality was seldom used. Finally, in quotation 3.3.3 it becomes clear that perceived transparency of clustering, i.e., every participant being able to trace activities, supported the pragmatic approach.

No.	Quotations concerning collaboration
Interactive transformation	
3.1.1	P3: 'I guess atmosphere fits with flair, no?' P1: 'Yes, sounds like it fits. Who wrote this? Did it have a different meaning or is it OK to connect these two?' P5: 'I wrote it. Let me check the other one, atmosphere. Where is it? Oh I have it. One second. Yes should be fine. […]'
3.1.2	P5: '[…]. **Oh, there is another one: Bavaria! It is also related to atmosphere.**'
3.1.3	P3: 'There is a lot around the topic flair / atmosphere. It seems like this is important.'
3.1.4	P7: 'I will put Bavaria next to flair and atmosphere. Is that alright?' P4: 'I also put beer festival next to it.'
3.1.5	P1 [connects text box no. 28 and 29] P2: 'Everyone is allowed to connect the text boxes, right?' P3: 'Sure, just have a look to see if you discover statements that fit well together.' P1 [connects text box no. 32 and 33] P2: 'Oh, OK.' P1 [connects text box no. 31 and 32] P2: 'The aim is to group the ideas. That makes it much easier for us to understand them.' P1 [connects text box no. 28 and 32] P1 [connects text box no. 36 and 39 unintended] P1: 'Oooooh, no!' P2: 'You can also disconnect them again, if you …' P1: 'Yes?' P4: 'Just drag it out again.' P1 [disconnects text box no. 36 and 39]
Contradiction transformation	
3.2.1	P2: 'A bicycle shop would be nice' P1: 'We could also include a shop for travel provisions' P2: 'We could also provide services around the bicycle, like a bicycle check. Some people would probably like their bicycle to be checked during their shopping at the shop next door.' P1: 'That sounds fantastic.' P5: 'Listen. It is already all there!' P1: 'No. Really?' P5: 'Yes. If you go to the main hall it is on the left side.' P2: 'Can't be!' P5: 'Yes. Believe me!' P1: 'OK. But tell me why I do not know it. I mean, I pass by this area every now and then.' P5: 'We have a flyer. It was also posted on the newsletter' P1: 'And how many people have used the service?' P5: 'Five, so far. It is not really frequently used, I have to admit.' P2: 'However, if we do not know, how should anyone else?' P3: 'Yes, you are right. We have to think about something else. Perhaps a newsletter is not the right communication channel or the service might also be too complex or we have to bundle it with something different.'
3.2.2	P3: 'I do not think we can do that. There are some limitations. I think changing the color has to be confirmed by the architect.' P2: 'Yes, changing the whole appearance has to be confirmed by the architect, but making some signs should be OK.' P3: 'It depends on what we want to do.'

	P1: 'OK, I have a suggestion: We think about the possibilities in the first place. Afterwards we cluster those concerning possible difficulties with the architect. Perhaps we can realize some of the solutions instantly. However, we have to ask our facility management. I know one of these guys.'
Pragmatic transformation	
3.3.1	P1: 'So, you are able to drag-and-drop the text boxes and then you can also group them. So, if we have too many ideas or ideas that are closely related … exactly, you are able to group them on top of each other, like you just did it. Consequently you have a group of ideas.' P2: 'You mean we can cluster, right?' [...] P1: 'And this is how you can also ungroup them. Just drag them out again.' [demonstrates this function]
3.3.2	P3: 'Whose idea is this? No. 37' P5: 'Could be mine, not sure. Perhaps someone reframed it. Did you [P1] write something to no. 37?' P1: 'Me? Where is it? Here it is. No, I don't think so.'
3.3.3	P4: 'Someone has put my idea to atmosphere.' P2: 'Yes, I guess that could have been me. Why? Is there something wrong?' P4: 'No, it is OK, but I think it is a different thing, because [long explanation]. So I think it rather refers to actual realization not to the general topic' P2: 'Oh, I understand. We could do it like this. Actually, I think you are right, but the idea has to be made clear. Shall I or do you want to do it.' P4: 'I'll do that. So I put it back to actual realization, right?'

Table 48: Quotations concerning collaboration

4.4   Outcomes

Besides generating a decent amount of innovations that provide fresh perspectives, outcomes of organizational innovation communities are used in four distinct manners: 'number of innovation ideas', 'scope of innovation ideas', 'igniting innovation projects', and 'latent innovation projects'.

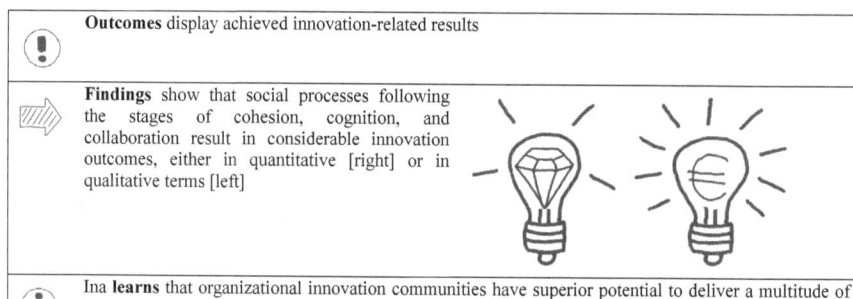

	Outcomes display achieved innovation-related results
	Findings show that social processes following the stages of cohesion, cognition, and collaboration result in considerable innovation outcomes, either in quantitative [right] or in qualitative terms [left]
	Ina **learns** that organizational innovation communities have superior potential to deliver a multitude of innovation outcomes

Table 49: Outcomes: Definition, findings, and learning

First, the *number of innovation ideas* relates to the number of innovation ideas that are created during hour-long sessions on the Open-I platform and more specifically on the collaborative whiteboard feature. The innovation ideas may be considered as first rough ideas for innovation concepts. These concepts are created during the collaboration stage but build on the explicated innovation ideas at the stage of cognition. Observations from many virtual whiteboard sessions indicate that an overwhelming number of ideas are created. For instance, in one session more than 100 ideas are created among six participants within one hour (quote 4.1.2). However, in other sessions at least 50 ideas are expressed, averaging 60 ideas (quote 4.1.1 and 4.1.3).

Second, the *scope of innovation ideas* in this case concerns the discontinuity of innovation ideas created within the organizational innovation community. It is shown that the community is able to create innovations that are new to the organization. However, the quality of such innovations varies: First, even though some of created innovations are new to the organizations they are easily realized as they do not need high investments and create revenues short-term (quotation 4.2.1 and 4.2.2). Second, some innovations are long term and need a considerable change in how the organization thinks about their business. These innovations may be understood as discontinuous innovations. Consequently, innovations created by the organizational innovation community include both immediately realizable innovations but also long-term innovations that need considerable adaptation of business models (quotation 4.2.3 and 4.2.4).

Third, *igniting innovation projects* refers to officially launched innovation projects within the hierarchical realms of the organizations. These project-based activities are realized in two major steps: 'presenting innovation concepts' and 'confirming innovation projects'. Foremost, quotation 4.3.1 describes the process in which developed innovation concepts are presented at a board meeting. In case 2, 23 innovation concepts are selected for presentation, as they reached a maturity level enabling executive board members to come to a decision. Finally, quotation 4.3.1 illustrates that nine of these innovation concepts were confirmed by board members to become official innovation projects, including project planning and budgeting. The number of innovations deployed in relation to time spent within the project exceeded all expectations and seems to be superior to other innovation methods currently applied, as quotation 4.2.4 emphasizes.

Fourth, *latent innovation projects* concern the transfer of results developed during activities within organizational innovation communities to daily work. This second-order construct refers to two first-order concepts: 'Being aware of solutions' and 'Applying solutions as occasions arise'. Primarily, quotation 4.4.1 illustrates that experiences concerning solutions are internalized and manifested within participants' frames of reference. Consequently, quotation 4.4.2 describes the transfer of developed solutions into daily work practices as soon as an occasion arises, in the sense that a project is redesigned and that experiences are included in this redesign.

No.	Quotations concerning outcomes
Number of innovations ideas	
4.1.1	Session 15 [76 innovation ideas]
4.1.2	Session 12 [105 innovation ideas]
4.1.3	Session 4 [52 innovation ideas]
Scope of innovations ideas	
4.2.1	P1: 'This idea, we can realize. We just have to make the contracts. That is easy.'
4.2.2	P3: 'The CEO presentation was fantastic. [...] nine innovation concepts **are to be implemented** by the [marketing] department. In half a year, the CEO wants to have a follow-up on these projects. [...] They really have to hurry now to get it implemented **instantly**, so that they are able to present some results in half a year.'
4.2.3	P3: 'Concerning the other ideas: We do not have the resources at the moment to realize them. I guess we still have to wait a bit, they are great but we have to slowly get into negotiations with our partners as it also affects them. This is a long-term decision. I think they will be convinced but we have to work on this for a bit.'
4.2.4	P3: 'I also did some lead user workshops. I mean, we have considerable experience with

	lead users. Also, one huge project that we are currently working on stemmed from one of these jobs. Still, we had to refine this idea quite considerably over recent years. But I am really astonished by the Open-I project. I mean, in comparison: **the time we need to prepare is less, we have all information documented** – normally we have to document everything, that takes ages and also important information gets lost, even though we take a lot of pictures and so on – and the **innovation concepts have really a quality** that is communicable to the CEO. Normally we have to prepare and refine the innovation ideas, which also takes a lot of time. **The results are also really good, I have to admit.**'
Igniting innovation projects	
4.3.1	P3: 'The CEO presentation was fantastic. [...] **nine innovation concepts** are to be implemented by the [marketing] department.'
4.3.2	P3: 'They really have to hurry now to get it implemented instantly, so that they are able to present some results in half a year.'
Latent innovation projects	
4.4.1	P5: 'I don't think that the innovation concept will make it. But anyway, I really like this idea and in approximately two to three years we have to redo it anyway. So, no matter if the concept will make it or not, in two or three years I will definitely include the basic idea, when we work on the topic anyway.'
4.4.2	P2: 'Thank you for the great ideas. I think we know way better now what the issues really are. I think some ideas are really freaky and I do not think that our customers would like them, but the underlying idea is great. So the next service conference is when, it starts in two months? That will be too tough but we have to start preparing the next one anyway in four months. So let's try to include some of the ideas at least.'

Table 50: Quotations concerning outcomes

5 Discussion

The categories identified from data analysis, as reported above, illustrate the process of innovation development in organizational innovation communities as they actually unfold. In the following, these categories are discussed against the theoretical background of community research. Further, wider implications of the categories are highlighted, several propositions suggested and a theoretical model presented.

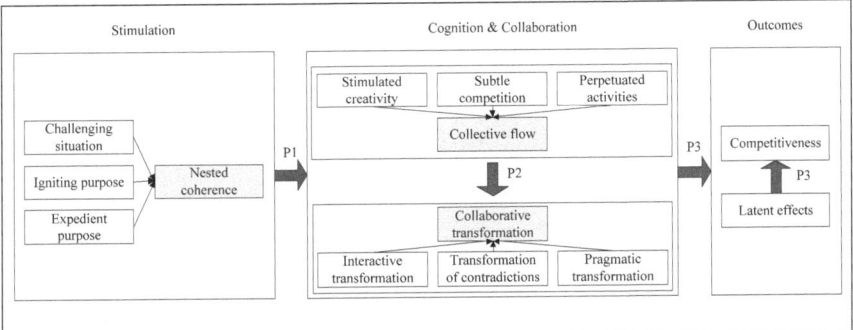

Figure 8: The social process of sensemaking for innovation

5.1   Nested coherence

Researchers at the intersection of sociology, organizational behavior, and innovation management emphasize a lacking understanding concerning the complex process of establishing coherence among weakly connected participants (Björk & Magnusson, 2009, p. 669). My findings, and particularly the notion of 'nested cohesion', explain the conditions under which cohesion among weakly connected participants, constrained by irregular, loose, and spontaneous interactions (Bogenrieder & Nooteboom, 2004, p. 295; Björk & Magnusson, 2009, p. 669), evolves. Based on data analysis, *nested cohesion* is defined as a collective state of mind, shared by multiple weakly connected participants providing a temporary impetus to exert extensive effort. Cohesion is nested because it emerges particularly around predefined purposes rather than towards the community at large. Following the data analysis, three means of nested cohesion – which are proposed to establish nested cohesion among participants – are identified: 1) igniting purpose, 2) challenging situation, and 3) useful purpose.

First, participants developed cohesion because they were working creatively on an igniting purpose. Interestingly, an igniting purpose was not bound to any particularly 'en vogue' topic or to the degree to which individuals had an interest in it, but rather to the way the purpose was communicated: participants had to be able to picture the purpose visually. With clear pictures in mind, participants gained fluid access to individual cognitive representations, such as prior experiences, understandings, etc. Consequently, an igniting purpose may serve as the sort of stimulation that facilitates participants to access a great variance of interpretations, as described in sensemaking theory.

Second, participants develop cohesion as they jointly master a challenging situation. For instance, the functionalities of the community platform challenged participants. However, instantly resolving these challenges creates an atmosphere among participants, colloquially described as 'sitting in the same boat'. Observations indicate that these feelings create the sort of social bonds among participants that are necessary for collective meaning creation and collective action, as described in sensemaking theory.

Third, cohesion among participants is developed as participants perceive their results as being valuable for the community or the organization at large. The data analysis supports the interpretation that perception of usefulness, i.e., perceptions of individual purpose-related effectiveness and high levels of purpose-related priority, supports effects of both igniting purpose and challenging situations, as their cognitive and social efforts within the community are of value for the organization.

Thus, cohesion sought by participants is nested in the combination of their perceptions of similar challenges for igniting a purpose and whether or not they perceive themselves as usefully contributing to innovation development. This formulation leads to the first propositions.

Proposition 1: *Members of the community will exert cohesion, if they are exposed to similar challenges, find the purpose igniting, and perceive their contribution as expedient. The community consequently exerts considerable effort to bring in individual interpretations.*

5.2   Collective flow

Community researchers widely agree that cognitive distance should receive more attention in scholarly work, especially as cognitively distant participants bring in variance in interpretations leading to higher-than-average discontinuous innovation (Amin & Roberts, 2008, p. 365; Hussler & Rondé, 2007, p. 300; Mørk et al., 2008, p. 12). The data analysis and notably the notion of 'collective flow' reveal that cognitively distant participants, i.e., participants with little overlap in cognitive frames, possess the abilities to fluidly reveal their divergent interpretations in an efficient way, fueling innovation development. Based on data analysis, *collective flow* is defined as a state of mind in which participants collectively immerse in an atmosphere of focus and effort while explicating immense quantities of ideas. Following the data analysis three major influences for collective flow to emerge are found: 1) stimulated creativity, 2) unconscious competition, and 3) perpetuated activities.

First, under conditions of stimulated creativity participants devise multiple individual interpretations at their own pace as they cognitively and naturally unfold, possibly stimulated by fellow participants' ideas. By means of explicating interpretations, participants accumulated ideas into repositories that provided the raw material for collaborative innovation development in posterior stages. Consequently, stimulated creativity enables cognitively distant participants to explicate their individual interpretations, serving as repositories for later collaborative innovation development.

Second, unconscious competition leads participants to increase their efforts to explicate as many interpretations as possible. As participants passively notice immense numbers of ideas popping up on the virtual whiteboard and hearing clattering noises from others' keyboards, participants seem to compete for high numbers of ideas. Consequently, unconscious competition increases explication of individual interpretations of cognitively distant participants, further fueling repositories for later collaborative innovation development. Hence, a competitive atmosphere inspires immense explication of interpretations.

Third, perpetuating activities is achieved by elongating efforts in conditions of waning concentration. For instance jokes deliver variation in communication and seem to enable access to additional cognitive interpretations related to a specific purpose. Hence, perpetuating activities elongate stimulated creativity, and unconscious competition leads to an increased level of explicating interpretations.

Thus, retaining divergent perspectives is kindled by stimulating creativity, subtle competition, and perpetuating efforts. All these three effects suggest that collective flow increases repositories of explicated individual interpretations. Collective flow is interpreted as a major influence on collaborative transformation as it provides an immense repository of individual interpretations. This formulation leads to the second proposition.

> **Proposition 2:** *The community will bring in divergent perspectives if it feels creatively stimulated, challenged, and if activities are perpetuated, resulting in huge repositories fueling collective meaning creation.*

5.3   Collaborative transformation

Community researchers widely agree that joint meaning creation is an unsettled question and deserves more attention, especially as it strengthens innovation development – e.g., as comprehensive understandings of a given issue are developed (Østerlund & Carlile, 2005, p. 101; Soekijad et al., 2004, p. 10; Palincsar et al., 1998, p. 10). Results indicate that constraints of joint meaning creation, especially participant hesitance, are resolved by 'collaborative transformation'. *Collaborative transformation* is defined as collectively combining multiple and divergent interpretations in recurrent and canonical trial and error cycles. Following the data analysis three key means of collaborative transformation are identified: 1) interactive transformation, 2) contradiction resolution, and 3) pragmatic transformation.

First, under conditions of interactive transformation participants analyze others' written interpretations and by this uncover clusters and interrelations. By means of organizing these interpretations visually, participants are able to refine and/or agree to the jointly created meaning. Hence, interactive transformation leads to comprehensively explicated innovations that capture divergent interpretations.

Second, contradiction resolution refers to collaborative identification of contradictions and their resolution. While interactively transforming interpretations, contradictions to widely accepted and applied paradigms became apparent to participants. However, as a comprehensive understanding was previously developed, they are able to resolve these contradictions and to use the contradictory situation to refine innovations even further.

Third, pragmatic transformation describes a hands-on process of innovation development in which participants aim at pragmatically developing the best possible solution. Disregarding possible political implications and personal offenses, participants are able to develop innovations efficiently. Hence, pragmatic transformation supports the efficient collaborative development of innovations.

Altogether, joint meaning creation is supported by interactive transformation, contradiction resolution, and pragmatic transformation. It enables participants to effectively develop innovations. This formulation leads to the third proposition.

Proposition 3: *The community will engage in collective meaning creation under conditions of paradigm identification, unbiased integration, and collaborative transformation, resulting in high amounts of innovation concepts ready for implementation.*

6 Conclusion

This part contributes to the understanding of collaborative and social processes in organizational innovation communities in the pursuit of innovation. Additionally, it provides a meaningful starting point to facilitate innovation development in organizational innovation communities. In other words, Ina profits from this part of the thesis by understanding how social processes unfold for innovation endeavors and by discovering possibilities to facilitate these processes.

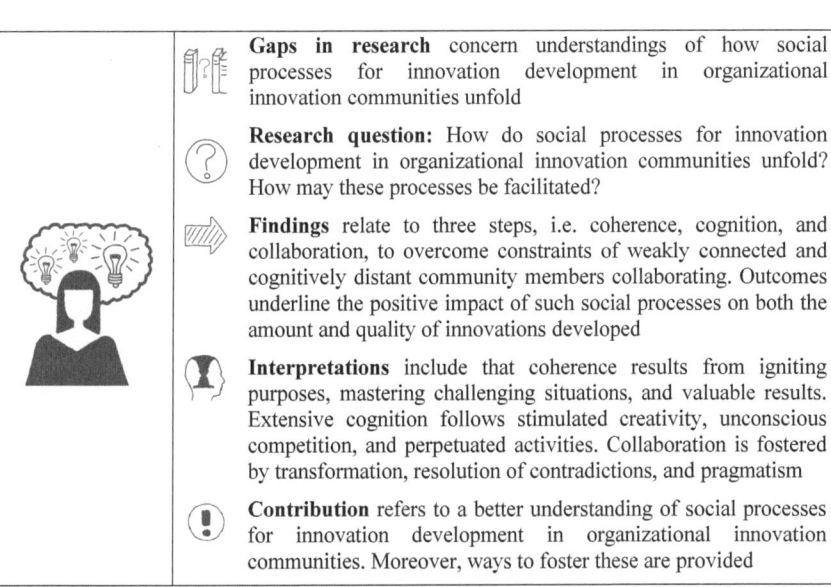

	Gaps in research concern understandings of how social processes for innovation development in organizational innovation communities unfold
	Research question: How do social processes for innovation development in organizational innovation communities unfold? How may these processes be facilitated?
	Findings relate to three steps, i.e. coherence, cognition, and collaboration, to overcome constraints of weakly connected and cognitively distant community members collaborating. Outcomes underline the positive impact of such social processes on both the amount and quality of innovations developed
	Interpretations include that coherence results from igniting purposes, mastering challenging situations, and valuable results. Extensive cognition follows stimulated creativity, unconscious competition, and perpetuated activities. Collaboration is fostered by transformation, resolution of contradictions, and pragmatism
	Contribution refers to a better understanding of social processes for innovation development in organizational innovation communities. Moreover, ways to foster these are provided

Table 51: Sensemaking: Summary of part IV

Based on a longitudinal action research approach in three field settings, a nuanced understanding of social processes for innovation pursuits is developed. It thereby shows that innovation development in such communities may be fostered to a large degree through the implementation of several strategies. Specifically, it is shown that coherence among weakly connected participants may be achieved quite easily, e.g., by briefly introducing a main but broadly defined topic. Additionally, truly collaborative innovation development, such as working on a virtual whiteboard, leads to collaborative flow experiences and consequently to high amounts of innovation ideas. Lastly, a collaborative transformation of first rough ideas into innovation concepts proved to be effective. In sum, this part demonstrates how the activities of participants may be directed to produce high amounts of innovations in an efficient way.

So far, the first and second major gaps in research, derived from reviewing community literature, have been explored. Now Ina is aware of how social processes unfold in organizational innovation communities for innovation pursuits and how she may be able to facilitate these processes. In the subsequent part, Norm's major struggle concerning what drives employees to engage in organizational innovation communities is put under the microscope.

Part V   Knowledge exchange in organizational innovation communities (empirical study III)

1 Setting the stage[16]

	Literature shows that motivation, trust, and status are crucial drivers for engagement. It has also shown how these factors may be implemented as an IT artifact
	Gaps in research refer to the use of more elaborate concepts and theories to understand the psychological functioning of employees in organizational innovation communities
	Norm learns that self-efficacy and positive affect induction positively influence the engagement of community members in terms of knowledge exchange

Table 52: Knowledge exchange: Existing literature, crucial gap, and learning

The preceding parts help Tom and Ina to cope with their struggles concerning the design and creation of supportive organizational contexts and understanding social processes for innovation pursuits. In the part at hand, Norm's struggle concerning major drivers for employees to engage in innovation endeavors is focused. One major means by which employees may engage in organizational innovation communities is knowledge exchange (Amin & Roberts, 2008; Breu & Hemingway, 2002). However, knowledge exchange activities as a pre-condition for innovation (Argote et al., 2000; Lin, 2007; MacDonald, 2008; Nonaka & Takeuchi, 1995) often remain limited due to stickiness of knowledge, cognitive distance among employees, and ambiguity of outcomes (Amin & Roberts, 2008; Brown & Duguid, 2001; Katz & Allen, 1982; Nooteboom, 2000; Østerlund & Carlile, 2005). In sum, these constraints reduce employees' motivation to engage actively. Thus, the present part examines the question *'what influences whether or not individuals engage in knowledge exchange in organizational innovation communities?'*

Although several factors may influence knowledge exchange in organizational innovation communities, the critical, and surprisingly understudied, role of cognitive and affective states is focused in this chapter. Specifically, it is argued that cognitive stimulation resulting in perseverance under conditions of difficulties and threats, ambition towards challenging tasks, and affective stimulation resulting in the availability of mental material, processable complexity, and cognitive flexibility, help to overcome main issues of knowledge exchange. With few valuable exceptions (Bishop, 2007; von Krogh et al., 2003) to date only limited insights on the causal relationship and extent to which cognitive and affective states influence knowledge exchange in organizational innovation communities are available.

To help facilitate this endeavor social cognitive theory (Bandura, 1986; Wood & Bandura, 1989) and behavioral motivation theory are applied (Gray, 1981, Gray, 1990; Ilies & Judge, 2005; Watson et al., 1999). *First*, social cognitive theory describes the relation between cognitive factors, external environments, and individual performance, e.g., the extent to which knowledge is exchanged (Bandura, 1986; Wood & Bandura, 1989). Wood and Bandura (1989) show that individual performance generally increases under conditions of

---

[16] Part V is based on a previous conference article presented and discussed (Bansemir, Neyer, & Möslein, 2011a) at the 2011 annual conference of the Academy of Management (AOM). The article is co-authored by Anne-Katrin Neyer and Kathrin M. Möslein An adapted version of this part is resubmitted to a well ranked journal after 'revise and resubmit' (Bansemir, Neyer, & Möslein). The lead author takes responsibility, e.g. for the design of the study, data gathering and analyses.

cognitive stimulation through self-efficacy induction. Thus, induced self-efficacy is proposed to enable cognitive stimulation, which leads to increased knowledge exchange among employees.

Second, behavioral motivation theory assumes that positive emotions, i.e., positive affect, reinforces pro-active initiatives, such as 'thinking outside-the-box', whereas negative emotions, i.e., negative affect, lead to avoidance behaviors such as resistance and escape (Gray, 1981, Gray, 1990; Ilies & Judge, 2005; Watson et al., 1999). Generally, positive affect is found to be fruitful in facilitating innovation-related knowledge exchange as it leads to cognitive variation that stimulates creativity (Amabile et al., 2005; Isen, 2001; Isen et al., 1992). Thus, inducing positive affect is proposed to enable the form of positive emotions, which encourages employees to exchange knowledge in organizational innovation communities.

By analyzing the influence of cognitive and affective states on knowledge exchange in organizational innovation communities, a more nuanced understanding of antecedents for successful innovation development is derived. Specifically, it is proposed that, on the basis of induced self-efficacy and positive affect individuals may be more or less likely to engage in knowledge exchange and thus to involve in innovation development. These propositions are tested using two experimental pre-test–post-test experiments.

2 Theoretical perspectives

In the following sections two hypotheses concerning the influence of self-efficacy and positive affect on knowledge exchange are derived. First, the possible influences of self-efficacy on knowledge exchange in organizational innovation communities are theoretically derived. Second, the influences of positive affect on knowledge exchange are theoretically derived. The derived hypotheses represent the theoretical foundation of the two conducted quantitative experimental studies presented subsequently. The following table displays key insights from theoretical considerations.

	Self-efficacy refers to self-beliefs to handle challenging situations. **Positive affect** may best be described as having positive feelings
	A gap in research exists concerning whether cognitive and affective states influence innovation-related activities, such as knowledge exchange, in organizational innovation communities and if so to what extent
	Key insights from social cognitive theory concerning self-efficacy indicate that increased self-efficacy [left] positively influences perseverance of activities in challenging situations [upper right], as well as under threatening conditions [middle right], and positively influences ambitions to tackle challenging tasks [lower right] 
	Research question: Does cognitive stimulation, such as self-efficacy induction, increase the innovation-related activities, such as knowledge exchange, of community members?
	A second gap in research relates to affective states and their influence on innovation-related activities, such as knowledge exchange in organizational innovation communities
	Key insights from behavioral motivation theory, and especially from affective states, hint at the potential of positive affect [left] to foster the availability of mental material [upper right], cognitive flexibility [middle right], and processible complexity [lower right] 
	Research question: Does affective stimulation, such as positive affect induction, increase the innovation-related activities, such as knowledge exchange, of community members?

Table 53: Theoretical background: Self-efficacy and positive affect and knowledge exchange

2.1   Self-efficacy and knowledge exchange

Researchers in psychology and innovation management have analyzed the relation among cognitive factors, external environments, and individual performance such as the extent of knowledge exchange (Bandura, 1986; Quigley et al., 2007; Wood & Bandura, 1989). They emphasize that cognitive stimulation delivers the needed impetus for community members to exchange knowledge in organizational innovation communities (Dahlander et al., 2008; Franke & Shah, 2003; Hsu et al., 2007; von Krogh et al., 2003; Lin, 2007; Quigley et al., 2007). Following social cognitive theory, three effects of self-efficacy, which are proposed to provide cognitive stimulation, are focused on: 1) perseverance under difficult conditions, 2) perseverance under threatening conditions, and 3) ambition towards challenging tasks.

First, self-efficacy exerts a positive influence on perseverance and the extent of effort in situations of difficulty, as individuals do not doubt their abilities (Wood & Bandura, 1989). Accordingly, the manner in which individuals cope with situations characterized by difficulties can be influenced by whether individuals exhibit more or less self-efficacy. Difficulties with regard to knowledge exchange often result from stickiness of knowledge, i.e., knowledge that is hard to transfer (Amin & Roberts, 2008; Hippel, 1994). For example, transfer of implicit knowledge remains difficult, as individuals find it demanding to codify implicit knowledge into clear descriptions or rules. Only if individuals are capable of overcoming these difficulties will knowledge be successfully exchanged. It is argued that if difficulties in knowledge exchange occur, those individuals who exhibit self-efficacy will exert greater effort to overcome these difficulties and will not quickly give up.

Second, self-efficacy is assumed to exert a positive influence on perseverance under threatening conditions. Wood and Bandura (1989) assume that threatening situations create high levels of stress and feelings of losing control. Self-efficacy reduces these feelings in threatening situations. In consequence, stress and perceived loss of control do not constrain the activities of individuals with self-efficacy in situations of threat. Studies show that knowledge exchange for innovation purposes can be classified as a complex and ambiguous process, in which results are rarely predictable (Østerlund & Carlile, 2005; Sawhney & Prandelli, 2000). Thus, employees with low self-efficacy are more likely to perceive the complex and ambiguous nature of knowledge exchange as threatening (Amin & Roberts, 2008; Soekijad et al., 2004). Hence, they are assumed to exert high levels of stress and feelings of losing control, which constrains knowledge exchange. In contrast, employees with high self-efficacy are less likely to get irritated and will thus engage in more knowledge exchange.

Third, self-efficacy stimulates individuals' ambition to take action on challenging tasks. Self-efficacy enables individuals to conceive difficult and challenging objectives and tasks as realistic options for action (Wood & Bandura, 1989). Consequently, individuals experiencing self-efficacy are likely not to focus on possible negative results of their engagement but to embrace challenging tasks with excitement. Still, exchanging knowledge in the pursuit of innovation may be challenging for employees who are not specifically educated or trained in this domain. This especially holds true when employees from various professional backgrounds are brought together. Indeed, studies show that collaborating with cognitively distant colleagues, e.g., employees from different professional backgrounds, challenges knowledge exchange. Such cognitively distant employees are bound in divergent worldviews, which often result in communication difficulties (Amin & Roberts, 2008; Brown & Duguid, 2001; Hussler & Rondé, 2007; Nooteboom, 2000). Therefore, individuals have to be perceptive to tackle occurring challenges. It is argued that employees with self-efficacy will be likely to meet these challenges, because they feel capable of handling them.

Although these three effects unfold and influence individuals' cognitions separately mainly by reducing psychological barriers, all propose that an increase of an employee's self-efficacy will lead to an increase of knowledge exchange in organizational innovation communities. Therefore the first hypothesis reads as following:

Hypothesis 1: *The more self-efficacy community members possess, the more likely they are to engage in knowledge exchange in organizational innovation communities.*

2.2   Affective states and knowledge exchange

Researchers in psychology and creativity have studied the relation between affective states and activities, such as knowledge exchange, profoundly (Amabile et al., 2005; Isen et al., 1992; Martin et al., 1993; Watson et al., 1999; Wood & Bandura, 1989). They find that positive affect induction facilitates innovation-related knowledge exchange as it leads to cognitive variation that stimulates creativity (Amabile et al., 2005; Isen, 2001; Isen et al., 1992). For instance, Fredrickson (1998, p. 304) states that "[...] experiences of certain positive emotions prompt individuals to discard time-tested or automatic (everyday) behavioral scripts and to pursue novel, creative, and often unscripted paths of thought and action". Three effects of positive affect which are proposed to provide positive emotions which stimulate knowledge exchange are identified: 1) the availability of mental material, 2) processable complexity, and 3) cognitive flexibility.

First, the availability of mental material, i.e., availability and access to potentially relevant information, is increased by positive affect. Several studies show that positive affect results in more information being accessible as individuals' scope of attention broadens (Amabile et al., 2005; Fredrickson, 1998; Isen et al., 1992; Martin et al., 1993). Accordingly, individuals' approaches towards accessing and combining mental material, i.e. their engagement in knowledge exchanging activities, can be influenced by their positive affect experiences. Thus, it is argued that the availability of mental material is increased under conditions of positive affect and hence reduces stickiness and fosters knowledge exchange.

Second, positive affect is found to enhance individuals' cognitive flexibility, as more and different associations are likely to emerge (Amabile et al., 2005; Isen et al., 1992). Isen et al. (1992) show that individuals under positive affect are open-minded, in the sense that they are unbiased, adaptive in thinking, embrace information, etc. Individuals are more likely "[...] to switch perspectives or entertain alternative perspectives to deal with data and solve a problem" (Isen et al., 1992, p. 58). Consequently, individuals experiencing positive affect are likely to embrace divergent information and adapt their thinking to increase cognitive flexibility. One major challenge for knowledge exchange results from cognitive distance among individuals. Individuals engaging in knowledge exchange need to react flexibly to information presented by cognitively distant individuals, even if that means that they have to change perspectives. Since individuals experiencing positive affect switch or entertain multiple perspectives, it is argued that a common understanding is more likely to emerge, forming the basis for knowledge exchange among cognitively distant individuals.

Third, research on affect posits that processable complexity is increased as a result of positive affect induction (Amabile et al., 2005). In this case, abilities to process complexity are elevated, as multiple factors are considered at the same time, realistic evaluations of situations are developed and beneficial coping is achieved (Isen, 2002). Knowledge exchange characterized by multiple social relations interfering, unclear procedures to follow, and rarely predictable results (Amin & Roberts, 2008; Østerlund & Carlile, 2005; Sawhney & Prandelli, 2000; Soekijad et al., 2004), is considered to be such a complex process. Individuals engaging in knowledge exchange need to be able to handle and make sense of the complexity arising from multiple influences. Given that the induction of positive affect supports the handling of complexity, it is argued that individuals experiencing positive affect will be able to process the inherent complexity, whereas individuals experiencing no positive affect will be less able to do so.

Although the three effects of affect unfold and influence individuals' cognitions differently mainly by providing additional cognitive resources, all propose that an increase of employees' positive affect experience should advance knowledge exchange in organizational innovation communities. Therefore, the second hypothesis reads as following:

Hypothesis 2: *The more positive affect individuals experience, the more likely they are to engage in knowledge exchange in organizational innovation communities.*

3 Research methods

Two experimental pre-test–post-test studies were conducted, testing each hypothesis in a university setting. Study 1 relates to hypothesis 1, in that self-efficacy beliefs were manipulated, i.e., the self-efficacy beliefs of participants were increased by encouragement. Study 2 tests hypothesis 2, in that positive affect is manipulated, i.e., positive affect was induced in participants by music and unexpected presents. Both studies were designed equally in the sense that they applied similar procedures, control techniques, measures, etc. Due to similarities, the design of the two experimental studies, the experimental task, and procedures are displayed first, followed by detailed explanations of the manipulations conducted. The following sections describe the research design, including experimental design, task and procedure, and manipulation and data collection.

3.1   Research design

A **pre-test–post-test between-subjects experimental design** is used (study 1: induced self-efficacy vs. no self-efficacy induction and study 2: induced positive affect vs. no positive affect induction). Both studies were conducted in lectures on business administration and included graduate students enrolled in a large German university. The first experimental study involved 78 participants with an average age of 24 years (standard deviation = 1.99), of whom 42 were male and 36 female. The second study involved 53 participants with an average age of 24 years (standard deviation = 3.20), of whom 36 were male and 17 female. Participants in both studies were randomly assigned to either experimental or control conditions. A self-reporting questionnaire concerning dependent variables was conducted at the beginning and at the end of the experiment, and manipulation checks were conducted at the end. Participation made up 10% of the overall class grading.

In both studies, participants completed an idea generation **task**, in which 50% of the participants experienced a stimulus. They were asked to develop strategies concerning the introduction of innovation communities. The specific objective of the task was to develop original but applicable ideas. The task was intentionally kept broad, in order to foster creating innovative views, ideas, and concepts. To develop innovative views, ideas and concepts, participants needed to have some basic impetus to find the task relevant for their studies. Two strategies were followed to ensure participation. First, a relatively small percentage (10%) of the overall class grade depended on the conducted activities. Second, community-related tasks were integrated in courses' tenors and provided a method to develop ideas for a written assignment that was due at the end of the course and made up a considerable percentage (50%) of the overall grade.

The two experimental studies followed similar **procedures,** as explained in the following. The Open-I platform was provided in both studies. To fulfill the task, participants were asked to traverse a three-step process, i.e., innovation generation, refinement, and evaluation. These procedures are in line with innovation development for instance at Cisco, Daimler, IBM, Intel, Siemens, and Vodafone among others (for a compilation of examples, see Zerfaß & Möslein, 2009). In the first step, participants generate a considerable number of minor and unstructured ideas, using the virtual whiteboard integrated in the community platform. As starting points, participants merge and refine these ideas in a second step, developing more sophisticated concepts. This process reflects the pilot organization's innovation development cycles as presented in part IV. In a third step, participants evaluate the concepts of other participants by applying a collaborative scoring method. Each step is announced within a lecture and participants had the following week, until the next session, to get the task done. This approach reproduces the introduction of the Open-I platform as it was experienced in the pilot organizations. The three lectures referred to the three workshops within the pilot organizations to produce a realistic setting comparable to the three pilots. Participants are free to choose the time and space for their activities in the community. Additionally, they were encouraged to generate ideas and refine or comment on others' concepts, as this is partly reflected in the grading.

In both studies, all participants were randomly and equally assigned to either the experimental or control group. Experimental and control groups were run simultaneously, however, working on separate but similar instances of the Open-I platform. Participants were told that two separate platforms were used due to technical issues to reduce participants' speculation about why two groups were implemented. Randomization ensured control of extraneous variables (Campbell et al., 1963; Christensen, 2007), whereas parallel groups provided control over relevant disturbing factors, because both groups were exposed to similar events (Sarris, 1990), e.g., information given during courses or other events not related to the university. Separation of the community platform for experimental and control groups was necessary to prevent spill-over effects of manipulations and thus ensured precise attribution of set stimuli.

3.2   Manipulations and data collection

To **manipulate self-efficacy** Wood and Bandura's (1989) work receives considerable attention. They distinguish between four major sources for individuals to enhance self-efficacy. Self-efficacy arises from 1) persuasion, i.e., realistic encouragement, 2) modeling, i.e., watching others succeeding through persistent effort, 3) mastery experiences, i.e., overcoming difficulties through persistent effort, and 4) fitness, i.e., enhanced physical status. In this study, the persuasion aspect was manipulated by creating encouraging messages on the Open-I platform concerning the given innovation task. In contrast to the implied importance of individuals as effective efficacy builders (Wood & Bandura, 1989), self-efficacy beliefs were manipulated by positioning encouraging messages directly on the community platform with no individual interaction whatsoever. Each time participants of the experimental group logged in to the community platform, the encouraging message was displayed for 15 seconds. The message was derived from existing literature on inducing self-efficacy beliefs based on persuasion (Bishop, 2007; Schwarzer & Jerusalem, 1995; Wood & Bandura, 1989). It states: 'You can achieve something exceptional! Start here and today: create something great by starting small, think outside regular patterns of thought and show that you are special! Search and create a gorgeous idea, together with your friends and student colleagues!' A message with a similar appearance, stating 'You are logged in', was posted on the community platform of the control group.

The design of stimuli for the **positive affect manipulation** was considerably influenced by Isen, Niedenthal and Cantor. Isen, Niedenthal and Cantor (1992) and Amabile et al. (2005) who distinguish three major strategies for inducing positive affect. Positive affect develops from 1) events, e.g., being surprised with an unexpected treat or gift, 2) film clips, e.g., showing a comedy, or 3) music, e.g., playing affect-laden music. The event and music aspects are manipulated within the experimental group. Showing film clips was not applied in order to keep manipulations as straightforward and short as possible. First, a virtual animated gift card, thanking participants for their engagement, was provided. Second, each time participants opened the community platform in their browser, energizing music started to play before fading out after around 30 seconds. The message printed on the gift card was derived from existing literature (Isen et al., 1992). It reads as following: 'Before you start to innovate on the community platform, there is one more thing: We, the members of the research team, are expressing our appreciation of your participation with this thank you card. So, this is for you!' Similar to study 1, a message stating 'You are logged-in' was posted on the community platform of the control group.

To **collect data** for the pre-test–post-test experimental studies, one week before launching the community platform and directly after completing assigned tasks, participants responded to a self-reporting questionnaire, including demographic information and questions concerning knowledge exchange. This approach was in accordance with a pre- and post-response measure, diminishing test effects, such as history and maturation effects, as they occurred in the experimental and control group alike and, thus, are controlled (Sarris, 1990; Stanley & Campbell, 1966).

To assess the dependent variable knowledge exchange two independent measures were assessed, i.e., 1) a self-reporting knowledge exchange scale and 2) a platform-based indicator of actual knowledge exchange activities. This strategy is followed to eliminate common method bias as a competing explanation for differences of variance, possibly resulting from self-reporting constructs (Avolio et al., 1991; Cote & Buckley, 1987; Donaldson & Grant-Vallone, 2002; Podsakoff et al., 2003).

First, knowledge exchange is assessed based on Zárraga and Bonache's (2005) five-item self-reporting construct, measured on a seven-point Likert scale (ranging from 1 = strongly disagree to 7 = strongly agree). The scale was adapted to the community context, e.g. the term 'work team' was replaced by 'community'. One sample item is: 'In the community, I have shared knowledge and experiences from my past (education or practice experience) that only I knew' (Zárraga & Bonache, 2005, p. 676). Similar to the findings of Zárraga and Bonache (*alpha*= 0.74) (2005: 669), acceptable reliability was found (*alpha* = 0.72).

Second, a platform-based knowledge exchange indicator is developed. The indicator automatically calculates points based on individuals' actual activities on the community platform. Participants gain points from three major knowledge-exchange activities: exchanging opinions, sharing ideas, and commenting on concepts. To ensure that knowledge was actually exchanged, only those activities rated useful by other participants increased the number of points. This additional measure served two specific objectives. First, due to differences in construction between the self-reported and the platform-based constructs, common method bias was restricted in terms of applying dissimilar types of constructs (Avolio et al., 1991; Donaldson & Grant-Vallone, 2002). Second, as data is collected via self-reported scales and actual activities on the community platform, common method bias is further limited in the sense that data are collected from different sources, i.e., self-reported vs. actual activities, (Cote & Buckley, 1987).

After completing assigned tasks and answering the post-response measure, the **manipulation check** is conducted at the beginning of the following lecture to minimize risks of priming and reduce response bias. The main page of the community platform and the screen after logging-in are shown. The box, which contained the message during the experiment, has been shown in plain color without any text. Participants were asked to remember the log-in situation and to answer a short questionnaire, which contained manipulation checks.

Study 1: To assess self-efficacy, Schwarzer and Jerusalem's (1995) ten items construct on self-efficacy, measured on a four point scale (ranging from 1 = not at all true to 4 = exactly true), were applied. One sample item reads as following: 'I can always solve difficult problems if I try hard enough'.

Study 2: To check manipulation of positive affect, positive affect is measured by applying Amabile et al.'s (2005, p. 379) self-rated mood measure, consisting of six items. The first five items relate to specific feelings, rated on a seven-point Likert scale (ranging from 1 = strongly disagree to 7 = strongly agree). One sample item reads as following: 'While working on the platform, I felt … … happy'. The sixth item relates to feelings of working with the community platform. It is measured on a seven-point Likert scale, ranging from 'extremely negative' to 'extremely positive'.

4 Findings

Before conducting actual statistical procedures concerning manipulation checks or comparisons of experimental and control groups, the study investigated whether the constructs achieved previously reported quality levels to ensure good item and scale quality in the present studies. Several tests were administered, but due to space constraints and predominantly satisfying results, only values for Cronbach's α are reported. The following two sections present the findings of the studies. First, results from the self-efficacy manipulation are displayed, followed by findings from the positive affect manipulation.

4.1 Manipulation checks

For **study 1** reliability coefficients were found to achieve acceptable values, e.g., Cronbach's α = 0.74). An ANOVA conducted on the manipulation check measures revealed significantly higher values of self-efficacy among participants of the experimental group compared to the control group (F [72] = 11.78, p < 0.01, adjusted R^2 = 0.23). In line with the manipulation, the analysis revealed lower mean values for self-efficacy for the control group (mean value = 2.81, standard deviation = 0.76) compared to the experimental group (mean value = 3.51, standard deviation = 0.73). These results support the belief that manipulation of participants' self-efficacy beliefs using platform-provided messages has been successful.

In **study 2** reliability coefficients of the positive affect construct produced acceptable values (Cronbach's α = 0.80). An ANOVA conducted on the manipulation check exhibited significantly higher values of positive affect among participants of the experimental group, compared to the control group (F [56] = 3.79, p < 0.1, adjusted R^2 = 0.05). Furthermore, descriptive statistics show increased mean values of the experimental group (mean value = 4.18, standard deviation = 0.75) compared to the control group (mean value = 3.65, standard deviation = 1.22). To conclude, results suggest that inducing positive affect, playing music and giving virtual presents, was successful and thus affirms the belief that the manipulations had the intended effects.

4.2   Test of hypotheses

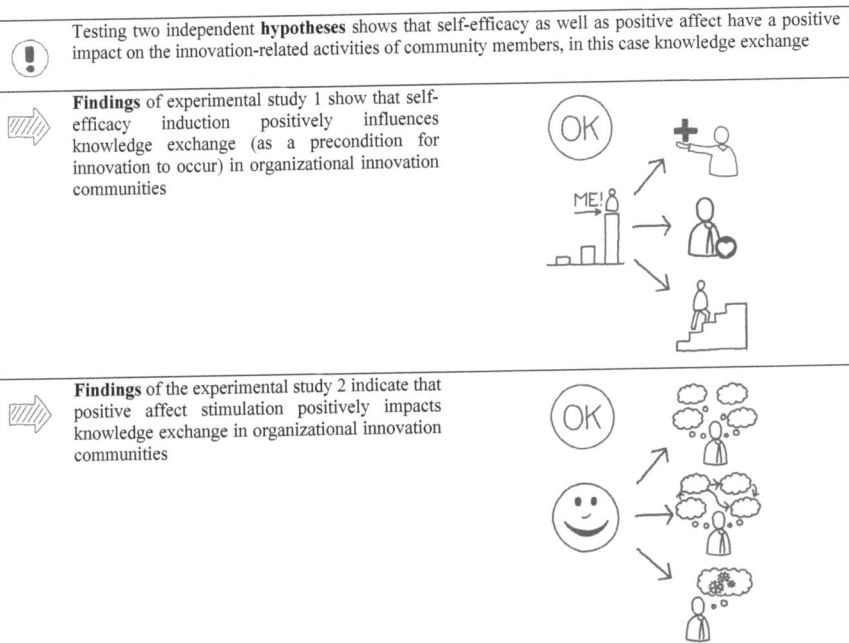

Testing two independent **hypotheses** shows that self-efficacy as well as positive affect have a positive impact on the innovation-related activities of community members, in this case knowledge exchange

Findings of experimental study 1 show that self-efficacy induction positively influences knowledge exchange (as a precondition for innovation to occur) in organizational innovation communities

Findings of the experimental study 2 indicate that positive affect stimulation positively impacts knowledge exchange in organizational innovation communities

Norm **learns** that believing in his own capabilities is crucial to be an effective contributor in organizational innovation communities. Additionally, he **learns** that positive affect (positive feelings) stimulates community members to contribute

Table 54: Self-efficacy and positive effect: Positive influence on knowledge exchange

The following section displays results from **study 1**, testing hypothesis 1. According to hypothesis 1, a positive relationship between self-efficacy and knowledge exchange is expected. The pre-test–post-test comparison reveals a highly significant main effect (F [156] = 9.530, p < 0.01, adjusted R^2 = 0.11), such that the experimental group with induced self-efficacy increased knowledge exchange by 0.80 points (pure effect, controlled for learning effects and influences of the community platform). Besides the fact that knowledge exchange increased in both groups, the experimental group achieved considerably higher values (control group: mean value = 4.36, standard deviation = 0.91, experimental group: mean value = 4.86, standard deviation = 0.76). Differences in mean values of the pre-test measure of self-reported knowledge exchange (mean of control group = 4.30 with a standard deviation = 0.64, mean of experimental group = 4.00 with a standard deviation = 0.87) are not significant (F [78] = 1.61, p > 0.1). The self-reported level of knowledge exchange of the pre-test measure results from knowledge being exchanged prior to the experiment within the lecture. Reliability coefficients of the knowledge exchange construct produce acceptable values (Cronbach's α = 0.71). Figure 1 displays the increase of knowledge exchange for both the control and the experimental group on the self-reporting scale.

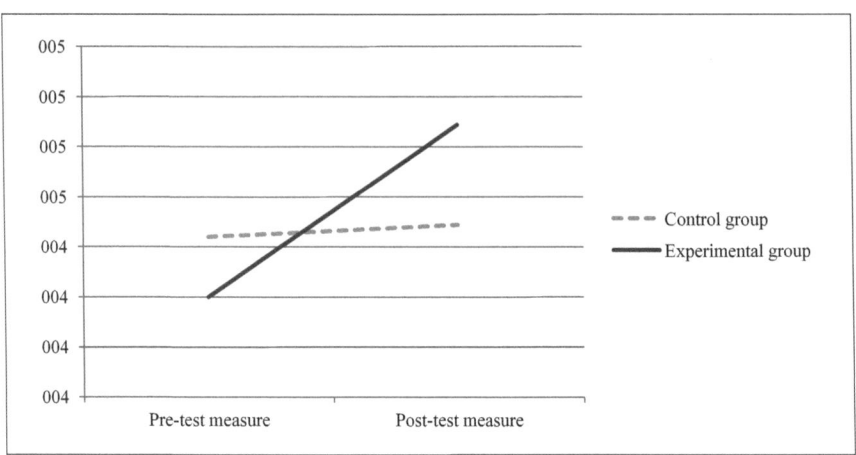

Figure 9: Influence of self-efficacy on knowledge exchange between control and experimental groups

In further support of hypothesis 1, results from the platform-based knowledge exchange indicator reveal that the experimental group conducted significantly more knowledge exchange activities compared to the control group (F [69] = 12.223, p < 0.01, adjusted R^2 = 0.14). Participants in the experimental group conducted nearly twice as many activities related to knowledge exchange as the control group (experimental group: mean value = 16.20, standard deviation = 10.79, control group: mean value = 8.79, standard deviation = 6.10). Figure 2 shows differences for the control and the experimental group on the platform-based indicator for knowledge exchange.

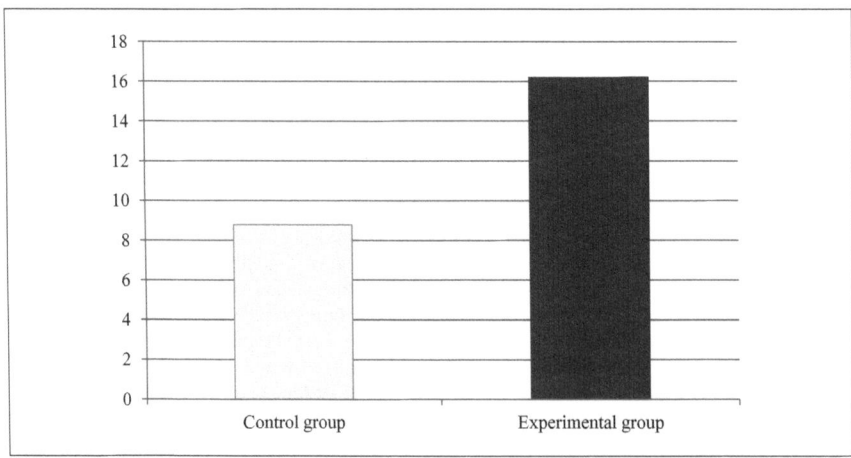

Figure 10: Influence of self-efficacy on knowledge exchange between control and experimental groups based on platform-based indicator of knowledge exchange

In line with the proposed theoretical arguments, results of the self-reported construct and the activities-based indicator support hypothesis 1. It is shown that induced self-efficacy via messages posted on the community platform increases the knowledge exchange of participants. Table 1 summarizes the results of the data analysis.

Time	Group	Measure	Mean value	Standard deviation
Pre-test	Control group	Self-reported	4.30	0.64
		Platform-based	n.a.	n.a.
	Experimental group	Self-reported	4.00	0.87
		Platform-based	n.a.	n.a.
Post-test	Control group	Self-reported	4.36	0.91
		Platform-based	8.79	6.10
	Experimental group	Self-reported	4.86	0.76
		Platform-based	16.20	10.79

Table 55: Summary of results of study 1 (self-reported knowledge exchange: $F [156] = 9.530$, $p < 0.01$, adjusted $R^2 = 0.11$; platform-based knowledge-exchange: $F [69] = 12.223$, $p < 0.01$, adjusted $R^2 = 0.14$)

In **study 2** the second hypothesis is tested. According to hypothesis 2, a positive relationship between positive affect and knowledge exchange is expected. The pre-test–post-test comparison reveals a highly significant main effect ($F [106] = 4.40$, $p < 0.05$, adjusted $R^2 = 0.07$), such that the experimental group with induced positive affect exhibits increased knowledge exchange by 0.84 points (pure effect, controlled for learning effects and influences of the community platform). Whereas knowledge exchange increased in the experimental group considerably, it slightly reduced among participants of the control group (Post-test measure: control group mean value = 4.23, standard deviation = 1.02, experimental group mean value = 4.35, standard deviation = 1.26, compared to pre-test measure: mean of control group = 4.24, standard deviation = 0.76, mean of experimental group = 3.52, standard deviation = 1.06). The reliability coefficients of the knowledge exchange construct produced acceptable values (Cronbach's α = 0.71). Differences in mean values of the pre-test measure of self-reported knowledge exchange were significant ($F [53] = 8.74$, $p < 0.01$, adjusted $R^2 = 0.12$), even though participants were randomly assigned to the groups. To exclude systematic bias due to differences in demographics – in terms of gender, age or experience distribution – between the experimental and control group additional statistical tests were run. Results confirmed that randomization was successful in terms of these variables and thus differences in pre-test ratings of knowledge exchange were not related to a systematic population bias. No evidence was found that differences in pre-test measures of knowledge exchange would negatively affect the explanatory power of the study. Figure 3 visualizes the above presented findings.

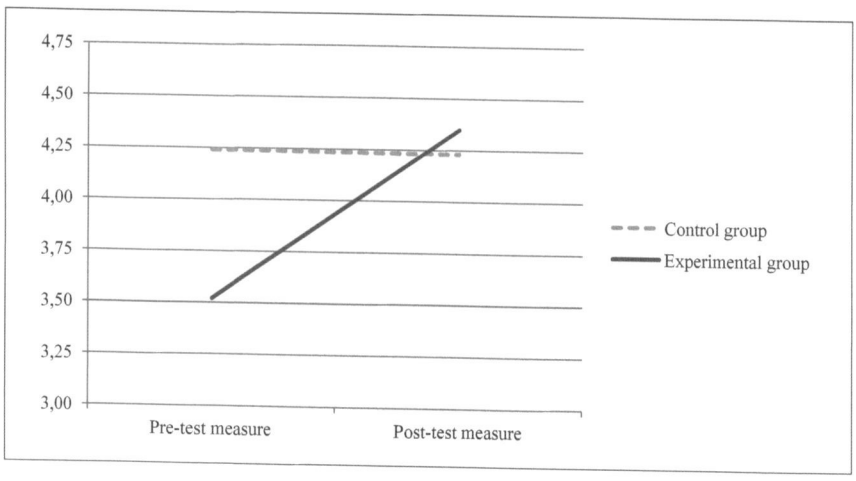

Figure 11: The influence of positive affect on knowledge exchange between control and experimental groups

Additionally, results derived from the platform-based indicator support hypothesis 2. They reveal a significant increase of knowledge exchange activities between the control and experimental group (F [69] = 4.41, p < 0.05, adjusted R^2 = 0.05). Participants increased knowledge exchange activities under treatment conditions by more than 50% (control group: mean value = 4.47, standard deviation = 3.74 compared to experimental group: mean value = 6.89, standard deviation = 5.60). Figure 12 summarizes these findings.

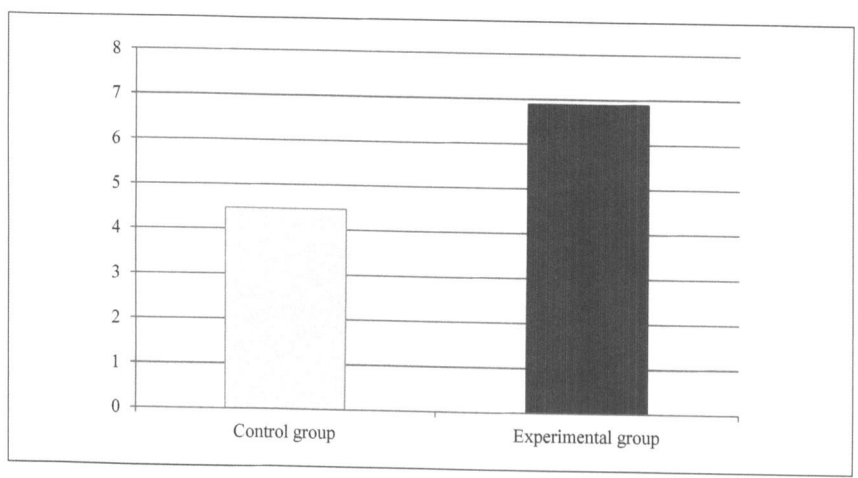

Figure 12: The influence of self-efficacy on knowledge exchange between control and experimental groups based on platform-based indicator of knowledge exchange

In sum, results of the self-reported construct and the activities-based indicator support hypothesis 2, as it is shown that induced positive affect via playing music and virtual presents posted on the community platform increased participants' knowledge exchange. In Table 2 key data are summarized.

Time	Group	Measure	Mean value	Standard deviation
Pre-test	Control	Self-reported	4.24	0.76
		Platform-based	n.a.	n.a.
	Experimental	Self-reported	3.52	1.06
		Platform-based	n.a.	n.a.
Post-test	Control	Self-reported	4.23	1.02
		Platform-based	4.47	3.74
	Experimental	Self-reported	4.35	1.26
		Platform-based	6.89	5.60

Table 56: Summary of results of study 2 (self-reported knowledge exchange: $F [106] = 4.40$, $p < 0.05$, adjusted $R^2 = 0.07$; platform-based knowledge exchange: $F [69] = 4.41$, $p < 0.05$, adjusted $R^2 = 0.05$)

5 Discussion

The studies extend previous research on the antecedents of knowledge exchange (Amin & Roberts, 2008; Breu & Hemingway, 2002) to the setting of organizational innovation communities. In general, findings reinforce the importance of research into individual factors impacting knowledge exchange in such communities. Additionally, underlying theoretical rationales are identified for why cognitive and affective states may affect knowledge exchange in the setting of the studies. We now turn to discussing in more detail the implications of key findings.

The presented findings have **theoretical implications**. *First*, cognitive stimulation, in the form of self-efficacy induction, is associated with increased knowledge exchange in organizational innovation communities. The above theorizing suggests that the effects of increased self-efficacy minimize frequent issues associated with knowledge exchange as follows. An increase in individuals' self-efficacy stimulates perseverance to overcome knowledge-exchange difficulties resulting from stickiness because individuals believe in their ability to do so. Also, they do not settle hastily with unsatisfactory results but exert persistence in order to overcome problematic situations. Additionally, it is possible that increased self-efficacy increases individuals' ambitions to take action in challenging tasks, such as engaging in knowledge exchange with cognitively distant individuals, as they feel capable of achieving positive results. Moreover, it is likely that increased self-efficacy fosters perseverance in knowledge exchange, characterized by ambiguity and complexity, as individuals perceive knowledge exchange as less stressful or uncontrollable compared to individuals with no increased self-efficacy.

In line with this, findings show that the effective stimulation of self-efficacy increases knowledge exchange in both measures, i.e., self-reported and actual behavior, significantly (both ps < 0.01) and predicts outcomes substantially (adjusted R^2s = 0.12 and 0.14 respectively). Moreover, the extent of knowledge exchange increases considerably, in terms of absolute measures (0.80 points self-rated measure, 100 % higher on behavior measure). Thus, it is suggested that the positive effects of increased self-efficacy, i.e., perseverance under conditions of difficulties or threats and ambition towards challenging tasks, are important antecedents to deal with the main issues of knowledge exchange within organizational innovation communities.

Second, positive emotions, in the form of induction of positive affect, are result in increased knowledge exchange. The above theorizing suggests that the effects of increased positive affect, i.e., availability of mental material, processable complexity, and cognitive flexibility, can explain the resulting increase of knowledge exchange because these common issues related to knowledge exchange are overcome. Increased positive affect among individuals enables access to and availability of mental material. This reduces stickiness, which often constrains knowledge exchange, as individuals find better ways to share knowledge with others. Furthermore, it is likely that increased positive affect results in increased cognitive flexibility. This results in a reduction of issues of incomprehensibility related to cognitive distance, as individuals appreciate divergent views and adapt to them. Likewise, it is possible that increased positive affect fosters individuals' ability to processable complexity. Hence, they are better equipped to cope with and make sense of the complexity and ambiguity inherent in knowledge exchange activities.

In line with this, the findings show that knowledge exchange among individuals is contingent upon the stimulation of positive affect. Both measures, self-reported and actual behavior, show that knowledge exchange under conditions of induced positive affect increase significantly (both ps < 0.05) and predict outcomes substantially (adjusted R^2s = 0.07 and 0.05 respectively). Additionally, the extent of knowledge exchange increases considerably, in absolute terms (0.80 points higher on self-rated measure, 50 % higher on behavior measure). Thus, it is suggested that the effects of positive affect, i.e., availability of mental material, processable complexity and cognitive flexibility, are important antecedents to deal with the main issues of knowledge exchange within organizational innovation communities.

Third, the findings suggest that even though the induction of self-efficacy and positive affect result in significantly increased knowledge exchange among individuals in the setting of the studies, there is an important difference in how these manipulations are reflected in the data set. Whereas individuals under conditions of induced self-efficacy are more likely to engage in knowledge exchange, the propensity of individuals who engage in knowledge exchange under conditions of induced positive affect is notably lower. Differences between self-efficacy and positive affect manipulations concerning knowledge exchange expressed in levels of significance (ps. of 0.01 and 0.05 respectively), extent of knowledge exchange (increase of behavior indicator of 100% compared to 50%) and predictive power (adjusted R^2s of > 0.12 compared to < 0.7) support this notion.

It is not appropriate to definitively explain why self-efficacy manipulations increased knowledge exchange notably more than positive affect manipulations, but the following reasonable explanation is offered. It is possible that existing cognitive barriers reduce the effectiveness of positive affect manipulations. As cognitive barriers are not removed through self-efficacy induction before manipulating positive affect in study 2, individuals might not believe in their skills to master difficulties or threatening situations and to tackle challenging tasks. In this case, the induction of positive affect, which is intended to result in elevated levels of mental material availability, processable complexity, and cognitive flexibility, might not unfold its full potential, as central cognitive barriers have not yet been removed. Thus, the intended outcome of positive affect induction, i.e. the activation of additional cognitive resources, might be restricted. It is concluded that positive affect only unfolds effectively if individuals feel capable of tackling a task. Therefore, one could argue that self-efficacy and positive affect should be induced in a sequential manner, in the sense that first self-efficacy may be increased and second positive affect.

Moreover, the findings suggest several preliminary **implications for practice**. A *first* practical implication of this finding for managers is that cognitive stimulation, such as self-efficacy induction, is crucial to foster knowledge exchange activities for innovation purposes in organizational innovation communities. For instance, the studies are able to show that self-efficacy induction alleviates cognitive barriers associated with stickiness, cognitive distance or ambiguity, as a precondition to increase innovation-related knowledge exchange. As managers aim at increasing knowledge exchange for innovation purposes, cognitive stimulation helps to overcome major barriers and hence should be an integral part of organizational innovation community management. This includes sensible monitoring of such cognitive states in qualitative as well as quantitative terms and the development of well-informed intervention strategies that stimulate cognitive states if needed.

A *second* practical implication refers to the important role of emotions for innovation-related knowledge exchange. For instance, the findings suggest that positive emotions, in the form of positive affect induction, extend cognitive resources resulting in increased knowledge exchange among individuals in organizational innovation communities. Thus when positive affect is induced, via affect-laden music and unexpected gifts, knowledge exchange is increased, as individuals access mental material more effectively, enhanced cognitive flexibility and increased processable complexity. Hence, results suggest that managers may also need to stimulate positive affect to enable cognitive variation as a major source for innovation. This includes careful monitoring of the community's atmosphere and timely intervention.

Third, the findings emphasize the importance of carefully managing individual factors on organizational innovation community platforms. Analyzing the data set reveals that the propensity of individuals who engage in knowledge exchange under conditions of induced positive affect is notably lower compared to conditions of induced self-efficacy. In this case, individuals may be constrained to fully exploit positive affect stimulations by cognitive barriers. To overcome these barriers, inducing self-efficacy may be a preliminary step for positive affect manipulations to unfold at the best. Hence, managers may need to stimulate cognitive and affective states in a sequential manner to unlock the full innovative potential of their employees.

In sum, both innovation and HR managers, who are eager to purposively foster innovation, should endeavor to monitor the cognitive and affective states of individuals to identify enablers and barriers for knowledge exchange within organizational innovation communities. Particularly, they should be more cognizant of the possible ways to stimulate individual factors, such as self-efficacy and positive affect, while designing innovation tools aiming at opening the innovation process within the organization.

Although the studies are a first step in considering cognitive and affective states that may influence knowledge exchange within organizational innovation communities, the results also suggest several **future fields for research**.

First, even though three effects for each induction type are proposed, i.e., self-efficacy and positive affect, which could lead employees to engage in knowledge exchange in organizational innovation communities, it is not feasible to directly test whether or not all effects were present in a similar intensity or in proposed directions, i.e., all effects having a positive influence on knowledge exchange. Indeed, these data are particularly hard to obtain, especially in experimental settings in which the number of feasible manipulations is limited due to needed effort. Further studies could continue to explore these important issues.

Second, there may also be a benefit to examining the extent to which distinct self-efficacy and positive affect manipulations affect different types of employees depending on their previous experiences concerning knowledge exchange in the context of innovation development within organizational innovation communities. One possibility could be that core inside innovators, i.e., innovators that are traditionally held responsible for innovation within R&D departments like Ina (Neyer et al., 2009), already possess high levels of self-efficacy with regard to their innovation capacity and, thus, will not respond with increased levels of knowledge exchange due to self-efficacy manipulations. In contrast, peripheral inside innovators, i.e. innovators across all business units not responsible for innovative activity by their job description like Norm (Neyer et al., 2009), might need a high level of self-efficacy induction to foster their engagement in knowledge exchange activities within such communities. Additionally, it is likely that employees who are engaging in organizational innovation communities for the first time may profit from self-efficacy manipulations to a higher extent, as they cannot rely on previous successes. In contrast, those employees who can build on previous expertise may not experience any differences. Further studies could continue to explore these important issues and help to shed light on how self-efficacy and positive affect manipulations influence distinct types of employees who have different levels of experience with regard to innovation development.

Finally, the use of an experiment in a university setting, including graduate students, may raise **limitations** about the external validity of the presented findings. However, Campbell and Stanley (1963) emphasize that external validity depends more on capturing all necessary dimensions than on the setting. Following this line, the studies do not reflect a laboratory experiment, but are designed as experiments that imitate actual work contexts with regard to engaging in organizational innovation communities as precisely as possible. It is important to mention that the design of tasks and procedures followed the experiences of the pilot studies within the three Open-I pilot organizations in great detail. As participants had to attend several other lectures, the work on the innovation task within the community platform required the students to prioritize their tasks, which corresponds to priority setting by employees within organizations. The impetus to participate in the experiment was comparable to the pilots within organizations. Also, as the community platform used for the experiment was the Open-I platform (as in the pilot organizations) the result is not distorted by means of the IT artifact. The step-wise development of innovations additionally reflects typical procedures for innovation development as employed in the pilot organizations the findings promise to generate high levels of external validity. However, despite the fact that the key dimensions imitated actual Open-I pilots, it is impossible to prove that all dimensions were captured accurately. Therefore, scholars are encouraged to conduct follow-up studies, testing self-efficacy and positive affect induction within organizational contexts. In particular, testing a sequential induction of self-efficacy and positive affect is proposed. This will allow a greater understanding of whether positive affect influences knowledge exchange in the same way as shown in the studies if cognitive barriers are removed beforehand.

6 Conclusion

This part contributes understanding of the engagement of employees in organizational innovation communities for innovation development. It refers to the third crucial research gap in organizational innovation community literature. Norm finds explanations as to why some employees engage more (or less) in organizational innovation communities. Strategies to foster greater employee engagement were also provided.

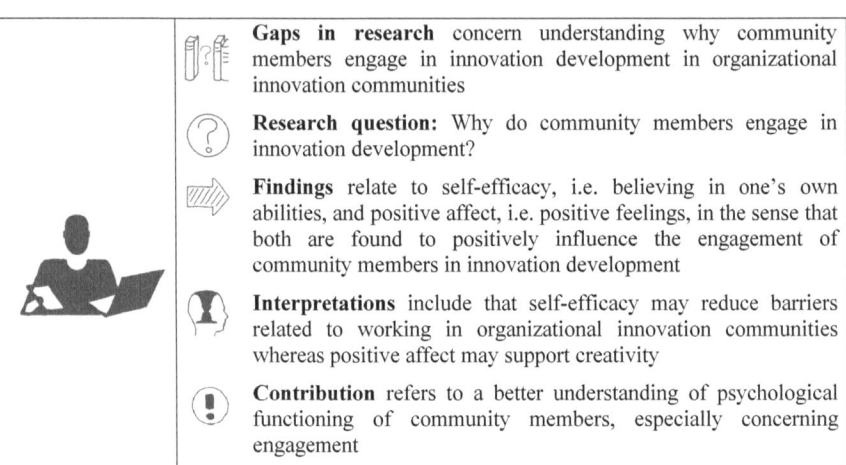

	Gaps in research concern understanding why community members engage in innovation development in organizational innovation communities
	Research question: Why do community members engage in innovation development?
	Findings relate to self-efficacy, i.e. believing in one's own abilities, and positive affect, i.e. positive feelings, in the sense that both are found to positively influence the engagement of community members in innovation development
	Interpretations include that self-efficacy may reduce barriers related to working in organizational innovation communities whereas positive affect may support creativity
	Contribution refers to a better understanding of psychological functioning of community members, especially concerning engagement

Table 57: Knowledge exchange: Summary of part V

Based on two quantitative experimental studies, Norm gains understanding of why community members engage more or less and how engagement may be fostered. It shows that stimulating self-efficacy and positive affect significantly increases community members' engagement and the development of innovations. Specifically, self-efficacy stimulation increases knowledge exchange as a pre-condition for innovation to occur significantly. Additionally, stimulating positive affect also leads to increased levels of knowledge exchange. Moreover, a technical stimulus intended to increase self-efficacy and positive affect by means of messages and music was successfully tested as a design element of community platforms (in this case the Open-I platform). Consequently, this part illuminates the engagement of community members in organizational innovation communities.

In the above, the third and final crucial gap identified in the existing literature was explored. Hence, Tom's, Ina's and Norm's major struggles have all been illuminated. The final part of this dissertation summarizes the findings and contributions of this thesis, derives practical as well as research implications, and ends with a brief epilogue.

Part VI Reflection

1 Summary and contribution

This thesis addresses organizational innovation communities, i.e. IT-supported communities within organizations in which community members engage in innovation development. Anticipating developments dating back to Linux's innovative approach of software development 20 years ago, this thesis adds another building block to turn this trend into a management innovation.

Even though many examples exist, including Cisco's I-Zone, Swarovski's I-flash, IBM's Innovation Jam, Daimler's Business Innovation Community, and Siemens's TechnoWeb, organizational innovation communities rarely unleash their full innovation potential. This is because of constraints that limit the efficient application of organizational innovation communities. Thus innovation communities may not be seen as management innovation, i.e. "[...] generation and implementation of a management practice, process, structure or technique that is new to the state of the art and is intended to further organizational goals" (Birkinshaw, 2008).

Overall, this thesis contributes to resolving crucial constraints and hence adds to establishing organizational innovation communities as a management innovation. This chapter delivers a summary of parts I to V, including the contributions made by its findings. In the subsequent chapter the practical implications of this thesis are displayed. Finally, directions for future research are provided.

Part I outlines the basic impetus of studying organizational innovation communities, i.e. to help turn organizational innovation communities into a management innovation. In doing so, it illustrates the potential of such communities to foster innovation pursuits. Moreover it also shows that organizational innovation communities are applicable in a wide set of organizations, whether they be (i) large or small or (ii) product or service organizations. Introducing Tom (top manager), Ina (innovation manager), and Norm ('normal' employee), crucial struggles that constrain organizational innovation communities are identified in narrative form. In total, Part I contributes to understanding the potential of community-based innovation development for organizations.

Part II delivers an analysis of contemporary literature on organizational communities. A systematic literature review method is chosen because extant literature has (i) inconsistent definitions, (ii) insufficient theoretical foundations, and (iii) unrelated strands of scholarly work dealing with organizational innovation communities. In sum, the literature review (i) provides a definition of organizational innovation communities, (ii) applies a theoretical framework, (iii) systematically organizes and summarizes extant research, and (iv) presents future avenues for research. First, based on a brief introduction of the historical development of community research, nine definitions of organizational communities are analyzed concerning similarities and differences. Juxtaposing these similarities and differences a definition of organizational innovation communities is derived. Second, to help organize unrelated and sometimes fragmented studies, a theoretical framework (i.e. the input, mediator, and outcome framework) is applied to the community context. Third, based on the theoretical framework 18 factors are identified and described. Fourth, avenues for future research are identified. These avenues are derived from analyzing crucial contents of organizational community literature. Finally, the systematic literature review concludes by identifying **three crucial research gaps**.

These are:

(1) Understanding the role of structural and cultural integration on innovation activities and outcomes

(2) Understanding social processes in the context of organizational innovation communities

(3) Understanding individual's engagement in organizational innovation communities

The following three empirical studies build on these critical gaps in research. Consequently, the literature analysis contributes to accumulative knowledge building, as it offers a detailed analysis of contemporary works. The following table displays crucial contents and gaps in literature related to Tom's, Ina's, and Norm's struggles.

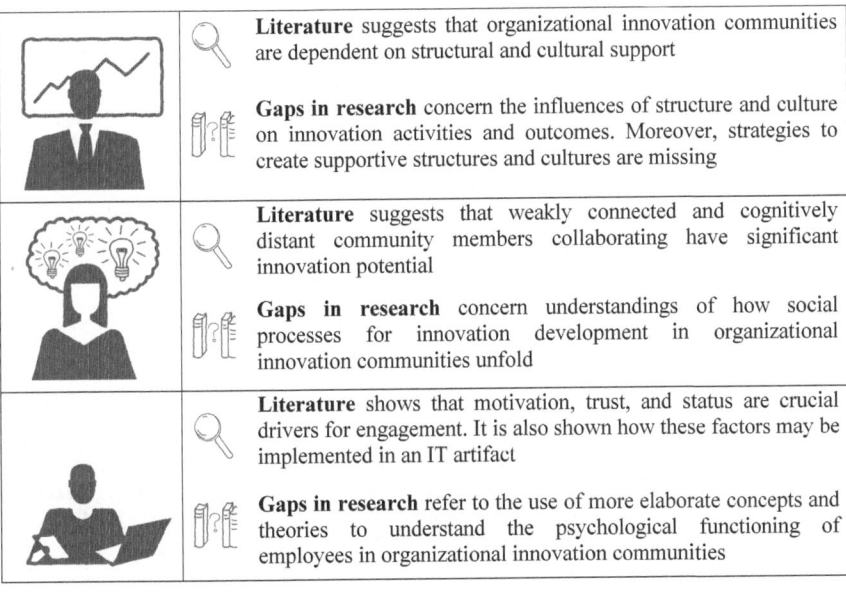

	Literature suggests that organizational innovation communities are dependent on structural and cultural support
	Gaps in research concern the influences of structure and culture on innovation activities and outcomes. Moreover, strategies to create supportive structures and cultures are missing
	Literature suggests that weakly connected and cognitively distant community members collaborating have significant innovation potential
	Gaps in research concern understandings of how social processes for innovation development in organizational innovation communities unfold
	Literature shows that motivation, trust, and status are crucial drivers for engagement. It is also shown how these factors may be implemented in an IT artifact
	Gaps in research refer to the use of more elaborate concepts and theories to understand the psychological functioning of employees in organizational innovation communities

Table 58: Summary of part II

Part III focuses on the organizational integration of organizational innovation communities by means of structural and cultural integration. To facilitate this endeavor, structuration theory is chosen as a theoretical perspective as it emphasizes (i) the role of organizational context to determine the activities and outcomes of employees and (ii) the role of employees in shaping these contexts, i.e. strategies to establish specific contexts. A multiple in-depth case study method is applied. In total 46 interviews were conducted in twelve organizations in heterogeneous settings. Three of the twelve studied organizations are pilots of Open-I and hence delivered additional longitudinal data. *Findings* relate to the identification of **four types of organizational integration**, resulting in varying levels of innovation activities and outcomes: *dyadic integration, cultural integration, structural integration*, and *no integration*. The data analysis suggests that cultural integration leads to high amounts of strategically relevant innovations while structural integration leads to the recurrent and pro-active delivery of innovations. Dyadic integration leads to a combination of the above mentioned (recurrent and pro-active delivery of high amounts of strategically relevant innovations) while no integration realizes none of the above mentioned. Additionally, three transition strategies are established to foster organizational integration: *initiation, negotiation, and narration*. *Initiation strategies* describe possibilities to support the emergence of organizational innovation communities. *Negotiation strategies* help to integrate organizational innovation communities either culturally or structurally. *Narration strategies* are a means to further organizational integration towards cultural *and* structural integration (dyadic integration). Based on literature emphasizing the important role of organizational integration (both cultural and structural), this part contributes to an understanding of how organizational contexts influence innovation activities and outcomes and how these supportive contexts may be established. Consequently, this part helps Tom understand how he can successfully anchor organizational innovation communities in his organization.

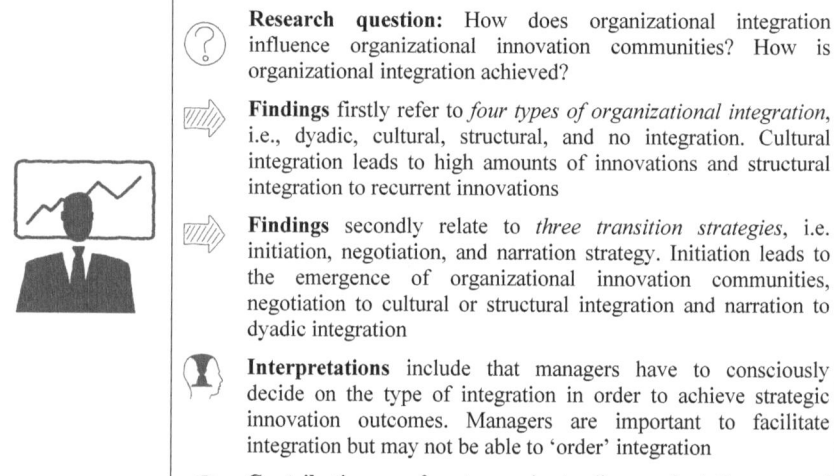

	Gaps in research concern the influences of structure and culture on innovation activities and outcomes. Moreover, strategies to create supportive structures and cultures are missing
	Research question: How does organizational integration influence organizational innovation communities? How is organizational integration achieved?
	Findings firstly refer to *four types of organizational integration*, i.e., dyadic, cultural, structural, and no integration. Cultural integration leads to high amounts of innovations and structural integration to recurrent innovations
	Findings secondly relate to *three transition strategies*, i.e. initiation, negotiation, and narration strategy. Initiation leads to the emergence of organizational innovation communities, negotiation to cultural or structural integration and narration to dyadic integration
	Interpretations include that managers have to consciously decide on the type of integration in order to achieve strategic innovation outcomes. Managers are important to facilitate integration but may not be able to 'order' integration
	Contributions refer to understandings of influences of organizational integration on outcomes and of strategies to achieve integration

Table 59: Summary of part III

Part IV focuses on social processes for innovation development in organizational innovation communities. Sensemaking theory helps to gain insight into these processes, as it delivers a framework to study how individuals collaboratively construct meaning. Action research methods are applied to study these social processes in three large service organizations over a time period of up to four years. This includes observations of actual innovation development taking place on the Open-I platform, which was a major data source. Results identify **three steps to overcome constraints:** *coherence, cognition,* and *collaboration*. First, coherence among community members is achieved if they master challenging situations jointly, have an igniting purpose, and perceive themselves as useful contributors. Meeting these criteria ultimately leads to nested coherence as a collective state consequently leading to extensive efforts. Second, cognition is stipulated if creativity of community members is stimulated, unconscious competition leads to extensive efforts and efforts are perpetuated, for instance by a good atmosphere. Consequently, collective flow as a shared state of mind immersing in an atmosphere of focus, energy and extensive effort is achieved. Third, collaboration builds on clear explications of community members' ideas, resolution of conflicting views and pragmatic clustering and combination of those ideas. Ultimately, community members participate in collaboratively transforming ideas into well-described innovation concepts.

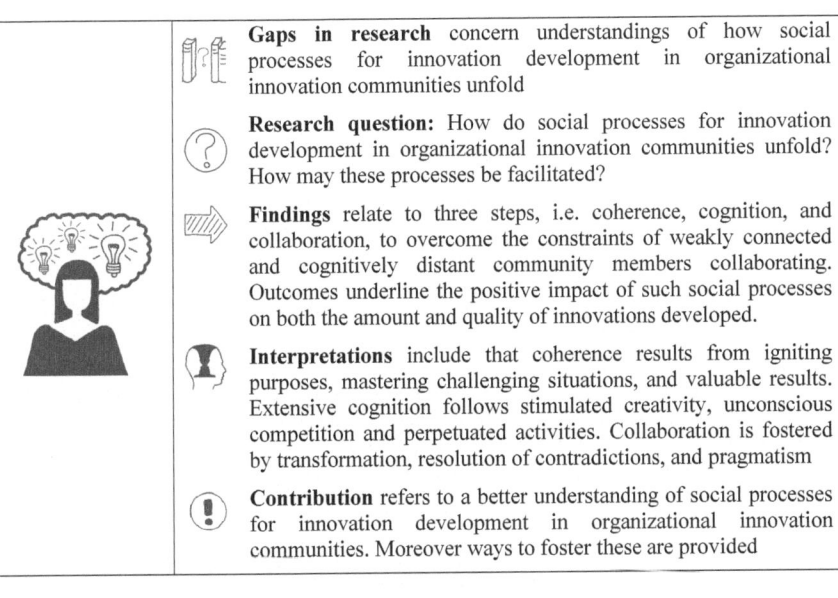

Gaps in research concern understandings of how social processes for innovation development in organizational innovation communities unfold

Research question: How do social processes for innovation development in organizational innovation communities unfold? How may these processes be facilitated?

Findings relate to three steps, i.e. coherence, cognition, and collaboration, to overcome the constraints of weakly connected and cognitively distant community members collaborating. Outcomes underline the positive impact of such social processes on both the amount and quality of innovations developed.

Interpretations include that coherence results from igniting purposes, mastering challenging situations, and valuable results. Extensive cognition follows stimulated creativity, unconscious competition and perpetuated activities. Collaboration is fostered by transformation, resolution of contradictions, and pragmatism

Contribution refers to a better understanding of social processes for innovation development in organizational innovation communities. Moreover ways to foster these are provided

Table 60: Summary of part IV

This part is able to demonstrate that such social processes lead to efficient innovation development and superior innovation outcomes. While literature emphasizes weakly connected and cognitively distant community members jointly creating meaning as a major source of innovation, research lacks a clear understanding about how processes in organizational innovation communities may be facilitated to leverage these potentials. Hence, this part adds to an understanding of how such social processes unfold in organizational innovation communities and how they may be facilitated. Ina gains considerable information about her role as an innovation manager to facilitate collaboration in organizational innovation communities.

Part V focuses on the engagement of community members in innovation related activities. It builds on social cognitive theory and behavioral motivation theory, in terms of self-efficacy and affective states, to explain engagement in knowledge exchange. Two explanatory and quantitative experimental pre-test–post-test studies using the Open-I platform are applied. It is important to mention that both self-efficacy and positive affect were stimulated by the platform, in the sense that messages were shown or music played, and hence this may be seen as a design element of the platform. Findings of this part reveal that (i) **self-**

efficacy stimulation and (ii) **positive affect** induction increase knowledge exchange among community members.

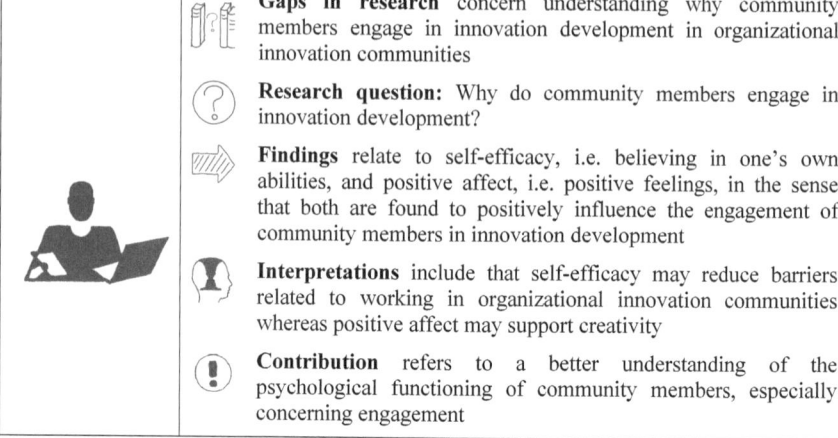

	Gaps in research concern understanding why community members engage in innovation development in organizational innovation communities
	Research question: Why do community members engage in innovation development?
	Findings relate to self-efficacy, i.e. believing in one's own abilities, and positive affect, i.e. positive feelings, in the sense that both are found to positively influence the engagement of community members in innovation development
	Interpretations include that self-efficacy may reduce barriers related to working in organizational innovation communities whereas positive affect may support creativity
	Contribution refers to a better understanding of the psychological functioning of community members, especially concerning engagement

Table 61: Summary of part V

Literature hints at the importance of stimulation – e.g. motivation by IT artifacts – however a clear understanding of the psychological functioning of employees in organizational innovation communities has been missing. This part contributes to a better understanding of this issue by applying elaborate concepts and theories from psychology. Consequently, Norm learns from this part why community members engage in organizational innovation communities.

Reflecting on **parts II, III, IV, and V** one might argue that each part – whether taken individually or as a whole – contributes to establishing organizational innovation communities as a management innovation. First, each part by itself extends current knowledge on organizational innovation communities. For instance, the literature review (part II) provides a detailed analysis of contemporary work in the context of organizational communities. Moreover, part III provides meaningful insights concerning the antecedents of organizational contexts and strategies to create supportive organizational contexts. Part IV adds understandings of social processes for efficient innovation development in organizational innovation communities. Lastly, part V delivers insights concerning community members' engagement. Second, regarding these parts as a whole adds additional value. One might argue that the engagement of employees is a precondition for any kind of innovation to occur in organizational innovation communities. Following this argument, fostering engagement by self-efficacy and positive affect (as described in part V) seems to be a crucial first step. Part IV may then be a necessary second step: the potentials of collaborative innovation development are fully leveraged to produce success stories, i.e. a narration strategy, to anchor organizational innovation communities culturally and structurally (as shown in part III). Summing up, in solving crucial research gaps identified from the literature review (part II) the empirical parts also feed into each other: Engagement of community members as a precondition for social processes (part V), social processes as a source for success stories (part IV), and success stories as a means to anchor organizational innovation communities (part III). In conclusion, each part – by itself but also as part of the whole – contributes to establishing organizational innovation communities as a management innovation.

Part VI summarizes the key contents of this thesis and describes its contributions. Moreover, in what follows implications for practice and avenues for future research are provided.

2 Practical implications

Managers attempting to apply organizational innovation communities as a means to foster innovation development within organizations may not want to count on risky and resource-consuming trial and error attempts with regard to exploiting such communities. Instead, they may want to rely on clear recipes to 'cook' the organizational innovation community according to their organization's strategic objectives. The practical implications of this thesis deliver such a recipe for managers, such as Tom or Ina, to configure organizational innovation communities according to their requirements. By extending the predominant emphasis of specific issues concerning organizational innovation communities towards a holistic and multi-level perspective, managers gain a more nuanced understanding of how to substantially increase the scale and scope of innovation development within organizations. Accordingly, the practical implications considered here address three major topics:

(1) **range of innovations**, i.e. amount and timely occurrence of innovations

(2) **degree of novelty**, i.e. achievement of discontinuous innovations

(3) **level of engagement**, i.e. participation as a prerequisite for innovations

Once managers like Tom or Ina are aware of the **range of innovations** they want to achieve while applying organizational innovation communities, they may identify the type of organizational innovation community they aim for. Answering two easy questions may help them to reach this goal: 1) how many innovations are strategically desired? and 2) how often are innovations strategically relevant?

The first question relates to the *amount of innovations* managers want to realize while building on organizational innovation communities. Findings clearly indicate that if managers aim at achieving high amounts of innovations, they have to integrate organizational innovation communities culturally. That is to say, managers have to encourage community members actively to develop innovations. For instance, CEOs have to specifically honor active community members by inviting them to present developed innovations in front of the executive board. Hence, if managers want to achieve high amounts of innovations, they have to culturally integrate organizational innovation communities.

The second question refers to the *timely occurrence of innovations* managers seek. Results show that if managers intend to generate innovations at high frequency, i.e. recurrently, they have to integrate organizational innovation communities structurally. That means that organizations have to integrate community activities and outcomes in incentive schemes, controlling metrics or have to provide resources for community activities. For instance, results show that providing a central budget to finance community-based innovation development leads to ongoing innovation development. Hence, if managers intend to accomplish high innovation frequency in organizational innovation communities, they have to structurally integrate such communities.

Summing up the last two paragraphs leads to the following conclusion, which is also supported by the empirical investigations of this thesis. If managers aim at generating *high amounts* of innovation at *high frequencies*, they have to integrate organizational innovation communities both structurally and culturally.

Three unique strategies are identified to achieve these predefined objectives: initiation, negotiation, and narration strategy. These strategies may be seen as steps to further innovation outcomes. First, the primary objective of *initiation strategy* is to launch organizational innovation communities at the outset. Once these communities are initiated managers may want to achieve high amounts of innovations or a high frequency of innovations. Findings indicate that applying the *negotiation strategy* helps managers to fulfill their chosen objective. Relying on this strategy enables managers to achieve high amounts of innovations if communities are culturally integrated. Alternatively, managers may apply this strategy to achieve high frequencies of innovation development if communities are structurally integrated. However, to achieve both innovation-related objectives simultaneously managers would profit from the *narration strategy*. The narration strategy builds on stories of successful innovation development and hence increases employee and top management awareness and support. Although each strategy might be applied separately to foster innovation outcomes, a step-wise use of these strategies systematically furthers organizational integration and innovation outcomes.

Once managers like Tom or Ina decide on the range of innovations, they may also be interested in how to increase the **degree of novelty** of innovations, i.e. achieving discontinuous innovations. Managers might want to apply the following three steps to fulfill this objective: establishing coherence, creating flow, and enabling collaboration.

The findings reveal that *establishing coherence* is a crucial first step for discontinuous innovation to occur. This is especially important if community members exhibit only irregular communication with each other, as they may not have enough motivation to collaborate. Managers therefore might profit from the finding that applying visual and narrative approaches to communicate innovation challenges has proven to be an effective strategy to create coherence.

Furthermore, *creating flow* experiences among community members while working on innovation tasks is a crucial second step for discontinuous innovation to occur. Flow leads to immersive quantities of high quality innovation ideas, particularly if community members experience this state of flow together, i.e. at the same time and at the same (virtual) place. The data show that this is facilitated by providing a virtual whiteboard in which every contribution is instantly visible to all other community members. Consequently, managers might be particularly interested in creating flow experiences among multiple community members as a source of discontinuous innovation.

Lastly, *enabling synchronous collaboration* of multiple community members is a decisive third step towards discontinuous innovation. It is shown that the synchronous collaboration of community members is especially fruitful to derive a comprehensive yet detailed understanding of discontinuous innovation concepts. Particularly, integrating the deviating perspectives of different community members has been shown to be crucial. Hence, managers might aim at enabling synchronous collaboration as a means to mature innovation concepts for later implementation.

Following these three steps, managers may have the capabilities to achieve high degrees of novelty when innovating in organizational innovation communities. Hence, besides increasing the scale of innovations developed, in terms of range of innovations, managers also gain knowledge concerning fostering the scope of innovations, in terms of the degree of novelty.

The last practical implication refers to the **levels of engagement** of employees as a crucial prerequisite to realize the desired scale and scope of innovations created by organizational innovation communities. Two strategies might be followed by managers, such as Tom and Ina, to increase the engagement of employees such as Norm: 1) overcoming barriers and 2) unleashing resources.

Findings indicate that cognitive stimulation is crucial to further the innovation-related activities of employees as major individual *barriers are overcome*. For instance, findings show that cognitive stimulation, in terms of self-efficacy induction, weakens individual barriers (e.g., resulting from ambiguity, stickiness or incomprehensibility). As managers seek to further innovation-related activities, cognitive stimulation may be a particularly fruitful strategy to overcome these barriers. Hence, managers should integrate cognitive stimulations as an essential part of community management, including sensible monitoring and the conscious design of intervention strategies.

Data also point to emotional stimulation as an effective strategy to foster the innovation-related activities of employees as individual mental *resources are unleashed*. For instance, it is shown that positive emotions extend the mental resources (e.g., availability of mental material, cognitive flexibility, and processable complexity) of community members. Managers that aim at fostering innovation-related activities might find emotional stimulation to be a particularly important strategy to unleash the existing mental resources of community members. Particularly, managers might be particularly sensible to the emotional states of community members and may derive strategies to create positive emotions instantly.

Altogether, practitioners – such as top managers like Tom, innovation managers like Ina or 'normal' employees like Norm – are provided with a toolbox to configure organizational innovation communities according to organizational objectives. The following figure illustrates levers concerning the range of innovations, the degree of novelty and the level of engagement. For instance, the left box shows that the range of innovations may be steered by the level of cultural and structural integration. It displays that the amount of innovations depends on cultural integration (left lever) whereas the frequency of innovations depends on structural integration (right lever). Based on the results of this dissertation, practitioners may be able to adapt organizational innovation communities to their specific needs.

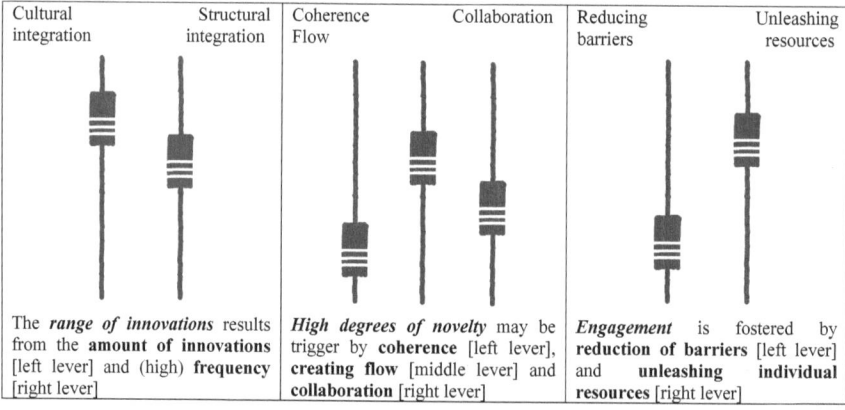

Cultural integration Structural integration	Coherence Collaboration Flow	Reducing Unleashing barriers resources
The *range of innovations* results from the **amount of innovations** [left lever] and (high) **frequency** [right lever]	*High degrees of novelty* may be trigger by **coherence** [left lever], **creating flow** [middle lever] and **collaboration** [right lever]	*Engagement* is fostered by **reduction of barriers** [left lever] and **unleashing individual resources** [right lever]

Table 62: Toolbox to configure organizational innovation communities

3 Research implications

It would be an exaggeration to claim that this thesis has filled all gaps in research. Several avenues of research remain untraveled. Moreover, this thesis also adds new avenues for future research that need to be answered. This chapter summarizes appealing ways to contribute to strengthening organizational innovation communities as a management innovation. To attain this aim research implications from the literature review and the three empirical studies are suggested. Additionally, overarching research implications are also provided.

The systematic literature review (**Part II**) offers a number of avenues for future research. It is important to mention that the literature analysis is based on an extensive number of scholarly works. The literature analysis was necessary because a sound and commonly accepted *theoretical foundation* is still missing. For instance, extant research has been fragmented into strands such as 'communities of practice', 'strategic communities', and 'online communities'. This is problematic as it remains unclear whether their findings are bound to specified contexts or can be generalized. The literature review may be seen as a first serious attempt to organize and summarize community literature using an input, mediator, and outcome framework. It may be worthwhile for researchers to follow up this avenue: extending and adapting the framework may contribute to cumulative knowledge building and may ultimately lead to a 'community theory'. But prior to that, researchers might be interested to clarify *cross-linkages* between factors, such as inputs, mediators, and outcomes. For instance, the influences of organizational contexts on outcomes have not been explicitly studied and therefore remain rather vague. Studying these cross-linkages would not only help to connect results but also to unearth exciting dependencies. Whilst part II provides multiple avenues for future research, three areas are identified as most crucial for further research: 1) *ICT features:* How do developments in Web 2.0 or Social Software affect communities?, 2) *Community dynamics*: How do changes in factors such as motivation and status unfold over time? and 3) *Possible interdependencies:* How are emergent states such as cognitive proximity, strength of ties and degree of centrality correlated with each other? In sum, a wide area for future theoretical or empirical research is provided.

Part III sheds light on the organizational integration of organizational innovation communities by means of structural and cultural integration. Traditionally, structural integration is accomplished by administrative procedures such as goal systems or performance appraisals; whereas cultural integration is mainly achieved through direct communication undertaken by managers in weekly departmental meetings or firm presentations. In other words, structural and cultural integration are still conducted within the hierarchical realms of the organization. However, it is not clear how to technically integrate these insights in organizational innovation community platform such as the Open-I platform. Findings show that integrating structural and cultural elements into community platforms may have the potential to dramatically increase employee engagement. For example, it has been identified that even a onetime participation by a CEO in a community can lead to high increase in the number of total participants and the quality of their participation. Hence, future research should focus on finding meaningful insights concerning the implementation of features that bring structural and cultural integration to the platform. Furthermore, data analysis reveals that the transition strategies are considerably driven by community members and hence may be seen as a 'bottom-up' process. Even though managers have been identified as playing a crucial role in fostering this process, it is not clear how they should go about it. Hence, future research may shed more light on the facilitating role of managers, especially concerning transition strategies.

Part IV derives considerable insights concerning social processes for innovation development in organizational innovation communities. Following the design science approach, some results have already been translated into IT artifacts such as the Massive Ideation software at Hyve and the Red Square software at SkyTec. Future research may look at these artifacts and their usage in detail to provide an even deeper understanding of social processes that are facilitated using such artifacts. Particularly, it might be of interest to explore IT *features* supporting each stage of social processes for innovation development, i.e. cohesion, cognition, and collaboration, in greater detail. Additionally, the data analysis discovers the concept of *'collective flow'*, i.e. a collective state of focus and effort. This concept is identified to have a positive impact on the number and quality of ideas to emerge. Even though a number of conditions that facilitate a state of collective flow are qualitatively identified, the exact relationship between them is not completely understood. Future research may perceive collective flow as a meaningful avenue to base more research on. The main aim of this part is to explore social processes in-depth. Therefore, studying selected cases longitudinally is a valid research strategy. But, as the sample of this study consists of firms from the service industry; one might speculate as to how generalizable the findings are (particularly transferring leanings to other industries). For instance, the mainly text-oriented collaboration feature may have successfully served the development of business concepts in service firms; other industries may need a different set of collaboration features (e.g. engineering firms may need collaboration features that rely more on visualizations). Hence, future research may consider replicating this study in other industries.

Part V explores the engagement of community members in innovation related activities. Using *theories from psychology* (social cognitive theory and behavioral motivation theory) this part explains different levels of member engagement. So far empirical investigations are mostly limited to the application of basic motivation concepts such as intrinsic and extrinsic motivation. However, applying more nuanced theories in the context of organizational innovation communities seem to have a lot of potential to further understanding psychological functioning of community members. Moreover, applying theories from psychology also has an impact on the *design of IT artifacts*. For instance, part V manipulates self-efficacy and positive affect using a technical stimulus, i.e. messages and music. Hence, psychological theories are a means to identify potential design elements for community platform such as the Open-I platform. Investigating other psychological constructs is likely to lead to even more design elements that foster innovation development within such communities. Hence, researchers may be curious to explore psychological theories and to identify candidates that may have significant impact on the design of such platforms.

An integrated perspective on **part II, III, IV, and V** reveals another avenue for future research. Even though this thesis studies different levels of analyses – i.e. organizational, community, and individual – it might not be considered as a *multi-level analysis*. It does not emphasize interactions between the levels analyzed. Whereas, this thesis hints at possible interactions among levels of analysis, future research may contribute to a better understanding of their interdependencies.

Overall, this thesis and related activities has already had some impact on research and practice. In *research* this work is contained in publications, conference schedules and research grants. So far the conducted research presented in this thesis has partly informed five book chapters[17] and led to three double-blind reviewed journal articles.[18] One article is currently on

---

[17] Bansemir, Habicht, Neyer, & Möslein, 2011; Möslein & Bansemir, 2008, Möslein & Bansemir, 2010; Möslein, Bansemir, & Haller, 2011; Möslein & Neyer, 2009.

resubmission after revision.[19] Additionally, the insights gained here from studying organizational innovation communities and related areas are frequently presented at national and international conferences.[20] Lastly, the results developed while working on the topic of the thesis has influenced large, currently granted, research proposals such as WiiPOD and Tandem. Consequently, further academic discussions concerning the topic of organizational innovation communities have been stimulated (at least in part) by this thesis.

To go beyond this thesis, the author aims to give other researchers *access to data and findings* related to this thesis. To achieve this objective, further publications are planned, based partly on data but also on the insights of this thesis. Publications may particularly refer to conference or journal publications to contribute to cumulative scholarly knowledge building in the area of organizational innovation communities. Moreover, collaboration between research and practice in the mindset of *design science* has proven to be a major success factor in providing meaningful results for this thesis. Even though the pressure to produce relevant results both for research and practice is indeed challenging, testing and experiencing derived solutions in 'real' settings has been extremely illuminating and motivating. Consequently, the author aims at further travelling this avenue. Practitioners as well as researchers are cordially invited to get involved in this journey.

---

[18] Bansemir & Neyer, 2009; Bansemir et al., 2012, Bansemir et al., 2011b.
[19] Bansemir et al.
[20] Exemplarily, contributions to selected national conferences in the year 2011 include Bansemir et al., 2011b; Bansemir, 2011a, Bansemir, 2011b; Bansemir et al., 2011a; Dul et al., 2011; Haller et al., 2011.

.

Reference list

Abdul-Rahman, A., & Hailes, S. (2000). Supporting trust in virtual communities. In *Proceedings of the 33rd Hawaii International Conference on System Sciences*. New York: IEEE Computer Society Press.

Adler, P. S. (2001). Market, hierarchy, and trust: The knowledge economy and the future of capitalism. *Organization Science, 12*(2), 215–234.

Agichtein, E., Liu, Y., & Bian, J. (2009). Modeling information-seeker satisfaction in community question answering. *ACM Transactions on Knowledge Discovery from Data, 3*(2), 1–27.

Alonzo, M., & Aiken, M. (2004). Flaming in electronic communication. *Decision Support Systems, 36*(3), 205–213.

Alstyne, M. van, & Brynjolfsson, E. (2005). Global village or cyber-balkans? Modeling and measuring the integration of electronic communities. *Management Science, 51*(6), 851–868.

Amabile, T. M., Barsade, S. G., Mueller, J. S., & Staw, B. M. (2005). Affect and creativity at work. *Administrative Science Quarterly, 50*(3), 367–403.

Amin, A., & Roberts, J. (2008). Knowing in action: Beyond communities of practice. *Research Policy, 37*(2), 353–369.

Ancona, D. G., & Caldwell, D. F. (1992). Demography and design: Predictors of new product team performance. *Organization Science, 3*(3), 321–341.

Andrews, D., Preece, J., & Turoff, M. (2002). A conceptual framework for demographic groups resistant to on-line community interaction. *International Journal of Electronic Commerce, 6*(3), 9–24.

Andrews, K. M., & Delahaye, B. L. (2000). Influences of knowledge process in organizational learning: The psychosocial filter. *Journal of Management Studies, 37*(6), 797–810.

Andriopoulos, C. (2001). Determinants of organisational creativity: A literature review. *Management Decision, 39*(10), 834–841.

Apostolou, D., Mentzas, G., Abecker, A., Maas, W., Georgolios, P., & Kafentzis, K. (2005). Challenges and directions in knowledge asset trading. *Intelligent Systems in Accounting, Finance & Management, 13*(1), 1–15.

Ardichvili, A., Page, V., & Wentling, T. (2003). Motivation and barriers to participation in virtual knowledge-sharing communities of practice. *Journal of Knowledge Management, 7*(1), 64–77.

Argote, L., Ingram, P., Levine, J. M., & Moreland, R. L. (2000). Knowledge transfer in organizations: Learning from the experience of others. *Organizational Behavior & Human Decision Processes, 82*(1), 1–8.

Arias, E. G., & Fischer, G. (Eds.). (2000). Boundary objects: Their role in articulating the task at hand and making information relevant to it. *International ICSP Symposium on Interactive and Collaborative Computing*.

Armstrong, A., & Hagel III, J. (1996). The real value of on-line communities. *Harvard Business Review, 74*(3), 134–141.

Armstrong, S. J., & Fukami, C. V. (2010). Self-assessment of knowledge: A cognitive learning or affective measure? Perspectives from the management learning and education community. *Academy of Management Learning & Education, 9*(2), 335–341.

Assimakopoulos, D., & Yan, J. (2006). The external linkages of community of practice: Integrating communities using internet technology forums. In H. Sherif & T. Khalil (Eds.), *Management of Technology: Selected Papers from the Thirteenth International Conference on Management of Technology* (pp. 199–221). Oxford: Elsevier.

Avolio, B. J., Yammarino, F. J., & Bass, B. M. (1991). Identifying common methods variance with data collected from a single source: An unresolved sticky issue. *Journal of Management, 17*(3), 571–588.

Ba, S. (2001). Establishing online trust through a community responsibility system. *Decision Support Systems, 31*(3), 323–336.

Bagozzi, R. P., & Dholakia, U. M. (2002). Intentional social action in virtual communities. *Journal of Interactive Marketing, 16*(2), 2–21.

Bagozzi, R. P., & Dholakia, U. M. (2006). Open Source software user communities: A study of participation in Linux user groups. *Management Science, 52*(7), 1099–1115.

Balasubramanian, S., & Mahajan, V. (2001). The economic leverage of the virtual community. *International Journal of Electronic Commerce, 5*(3), 103–138.

Balogun, J., & Johnson, G. (2004). Organizational restructuring and middle manager sensemaking. *Academy of Management Journal, 47*(4), 523–549.

Bandura, A. (1986). *Social Foundations of Action and Thought: A Social Cognitive View.* Englewood Cliffs, NJ: Prentice-Hall.

Bansemir, B. (2011a). Organizational innovation communities. In *Mensch und Computer. ÜberMedien ÜberMorgen* (pp. 1–4). Chemnitz: Gesellschaft für Informatik.

Bansemir, B. (2011b). Sensemaking in organizational innovation communities. In *Academy of Management (AOM)* (pp. 1–29). San Antonio.

Bansemir, B., Habicht, H., Neyer, A.-K., & Möslein, K. M. (2011). Open-I: Periphere interne Innovatoren erfolgreich integrieren. In H. Jakobsen & B. Schallock (Eds.), *Innovationsstrategien jenseits traditionellen Managements. Wissenschaftliche und praktische Ergebnisse des Förderschwerpunktes* (pp. 163–171). Stuttgart: Fraunhofer IRB.

Bansemir, B., & Neyer, A.-K. (2009). From idea management systems to interactive innovation management systems: Designing for interaction and knowledge exchange. In H. R. Hansen (Ed.): *Konzepte, Technologien, Anwendungen* (pp. 861–870). Vienna: Computer-Ges.

Bansemir, B., Neyer, A.-K., & Möslein, K. M. (2012). Knowledge exchange in intra-organizational innovation communities: The role of cognitive and affective states, *BuR – Business Research, 5*(1), 43–58.

Bansemir, B., Neyer, A.-K., & Möslein, K. M. (2009). Intra-organizational innovation communities: Towards a framework. In *European Academy of Management (EURAM)* (pp. 1–39). Rome.

Bansemir, B., Neyer, A.-K., & Möslein, K. M. (2010). Organizational communities. In *Research Colloquium Innovation and Value Creation* (pp. 1–48). Beilngries.

Bansemir, B., Neyer, A.-K., & Möslein, K. M. (2011a). Knowledge exchange in intra-organizational innovation communities: The role of cognitive and affective states. In *Academy of Management (AOM)* (pp. 1–32). San Antonio.

Bansemir, B., Neyer, A.-K., & Möslein, K. M. (2011b). Towards a taxonomy of corporate innovation communities. In *Informatik 2011 – Informatik schafft Communities* (pp. 1–6). Berlin: Gesellschaft für Informatik.

Bansemir, B., Neyer, A.-K., & Möslein, K. M. (2012). Anchoring corporate innovation communities in organizations: A taxonomy. *International Journal of Knowledge-Based Organizations, 2*(1), 1–20.

Barab, S. A., MaKinster, J. G., & Scheckler, R. (2003). Designing system dualities: Characterizing a web-supported professional development community. *Information Society, 19*, 237–256.

Barrett, M., Cappleman, S., Shoib, G., & Walsham, G. (2004). Learning in knowledge communities: Managing technology and context. *European Management Journal, 22*(1), 1–11.

Bartunek, J. M., Rousseau, D. M., Rudolph, J. W., & DePalma, J. A. (2006). On the receiving end: Sensemaking, emotion, and assessments of an organizational change initiated by others. *Journal of Applied Behavioral Science, 42*(2), 182–206.

Bauer, H. H., & Grether, M. (2005). Virtual community: Its contribution to customer relationships by providing social capital. *Journal of Relationship Marketing, 4*(1/2), 91–109.

Bechky, B. A. (2003). Sharing meaning across occupational communities: The transformation of understanding on a production floor. *Organization Science, 14*(3), 312–330.

Beenen, G., Ling, K., Wang, X., Chang, K. F. D., Resnick, P., & Kraut, R. E. (2004). Using social psychology to motivate contributions to online communities. *CSCW, 6*(3), 212–221.

Bergquist, M., & Ljungberg, J. (2001). The power of gifts: Organizing social relationships in open source communities. *Information Systems Journal, 11*(4), 305–320.

Bertels, H. M. J., Kleinschmidt, E. J., & Koen, P. A. (2011). Communities of practice versus organizational climate: Which one matters more to dispersed collaboration in the front end of innovation? *Journal of Product Innovation Management, 28*(5), 757–772.

Bettiol, M., & Sedita, S. R. (2011). The role of community of practice in developing creative industry projects. *International Journal of Project Management, 29*(4), 468–479.

Bieber, M., Engelbart, D., Furuta, R., Hiltz, S. R., Noll, J., Preece, J., et al. (2002). Toward virtual community knowledge evolution. *Journal of Management Information Systems, 18*(4), 11–35.

Birchall, D., & Giambona, G. (2007). SME manager development in virtual learning communities and the role of trust: A conceptual study. *Human Resource Development International, 10*(2), 187–202.

Birkinshaw, J., Hamel, G., & Mol, M. J. (2008). Management innovation. *The Academy of Management Review (AMR), 33*(4), 825–845.

Bishop, J. (2007). Increasing participation in online communities: A framework for human-computer interaction. *Computers in Human Behavior, 23*(4), 1881–1893.

Björk, J., & Magnusson, M. (2009). Where do good innovation ideas come from? Exploring the influence of network connectivity on innovation idea quality. *Journal of Product Innovation Management, 26*(6), 662–670.

Blanchard, A. L., & Markus, M. L. (2004). The experienced "sense" of a virtual community: Characteristics and processes. *SIGMIS Database, 35*(1), 64–79.

Boczkowski, P. J. (1999). Mutual shaping of users and technologies in a national virtual community. *Journal of Communication, 49*(2), 86108.

Bogenrieder, I., & Nooteboom, B. (2004). Learning groups: What types are there? A theoretical analysis and an empirical study in a consultancy firm. *Organization Studies, 25*(2), 287–313.

Bolton, G. E., Katok, E., & Ockenfels, A. (2004). How effective are electronic reputation mechanisms? An experimental investigation. *Management Science, 50*(11), 1587–1602.

Borzillo, S., Aznar, S., & Schmitt, A. (2011). A journey through communities of practice: How and why members move from the periphery to the core. *European Management Journal, 29*(1), 25–42.

Boschma, R. (2005). Proximity and innovation: A critical assessment. *Regional studies, 39*(1), 61–74.

Brailsford, T. W. (2001). Building a knowledge community at Hallmark Cards. *Research Technology Management, 44*(5), 18–25.

Brazelton, J., & Gorry, G. A. (2003). Creating a knowledge-sharing community: If you build it, will they come? *Communications of the ACM, 46*(2), 23–25.

Brentani, U. de, & Kleinschmidt, E. J. (2004). Corporate culture and commitment: Impact on performance of international new product development programs. *Journal of Product Innovation Management, 21*(5), 309–333.

Breu, K., & Hemingway, C. (2002). Collaborative processes and knowledge creation in communities-of-practice. *Creativity & Innovation Management, 11*(3), 147–153.

Brown, J., Broderick, A. J., & Lee, N. (2007). Word of mouth communication within online communities: Conceptualizing the online social network. *Journal of Interactive Marketing, 21*(3), 2–20.

Brown, J. S., & Duguid, P. (1991). Organizational learning and communities-of-practice: toward a unified view of working, learning, and innovating. *Organization Science, 2*(1), 40–57.

Brown, J. S., & Duguid, P. (2001). Knowledge and organization: A social-practice perspective. *Organization Science, 12*(2), 198–213.

Brown, N. R. (2002). "Community" metaphors online: A critical and rhetorical study concerning online groups. *Business Communication Quarterly, 65*(2), 92–100.

Bungart, S., & Köhler, K. (2009). Innovation durch Kommunikation und Kollaboration. In A. Zerfaß & K. M. Möslein (Eds.), *Kommunikation als Erfolgsfaktor im Innovationsmanagement: Strategien im Zeitalter der Open Innovation* (pp. 355–366). Wiesbaden: Gabler.

Burns, T., & Stalker, G. M. (1994). *The Management of Innovation.* New York: Oxford University Press.

Butler, B., Sproull, L., Kiesler, S., & Kraut, R. (2007). Community effort in online groups: Who does the work and why? (pp. 171-194). In S. Weisband. (Ed.), *Leadership at a distance.* NY: Lawrence Erlbaum Associates/Taylor Francis.

Campbell, D. T., Stanley, J. C., & Gage, N. L. (1963). *Experimental and quasi-experimental designs for research.* Chicago: Rand McNally.

Campbell, J., Fletcher, G., & Greenhill, A. (2007). Sustainable virtual world ecosystems. *SIGMIS Database, 38*(4), 29–31.

Capon, N., Farley, J. U., Lehmann, & Hulbert, J. M. (1992). Profiles of product innovators among large US manufacturers. *Management Science, 38*(2), 157–169.

Casalo, L. V., Flavian, C., & Guinaliu, M. (2008a). Fundamentals of trust management in the development of virtual communities. *Management Research News, 31*(5), 324–338.

Casalo, L. V., Flavian, C., & Guinaliu, M. (2008b). Promoting consumer participation in virtual brand communities: A new paradigm in branding strategy. *Journal of Marketing Communications, 14*(1), 19–36.

Case, S., Azarmi, N., Thint, M., & Takeshi, O. (2001). Enhancing e-communities with agent-based systems. *IEEE Computer, 34*(7), 64–69.

Castelfranchi, C., & Tan, Y.-h.(2002). The role of trust and deception in virtual societies. *International Journal of Electronic Commerce, 6*(3), 55–70.

Chan, C. M. L., Bhandar, M., Oh, L.-B., & Chan, H.-C.(2004). Recognition and participation in a virtual community. *Proceedings of the 37th Hawaii International Conference on System Sciences*. New York: IEEE Computer Society Press.

Chang, B., Cheng, N.-H., Deng, Y.-C., & Chan, T.-W. (2007). Environmental design for a structured network learning society. *Computers & Education, 48*(2), 234–249.

Chang, L.-J., Chou, C.-Y., Chen, Z.-H., & Chan, T.-W. (2004). An approach to assisting teachers in building physical and network hybrid community-based learning environments: The Taiwanese experience. *International Journal of Educational Development, 24*(4), 383–396.

Chang, L.-J., Yang, J.-C., Deng, Y.-C., & Chan, T.-W. (2003). EduXs: Multilayer educational services platforms. *Computers & Education, 41*(1), 1–18.

Cheliotis, G. (2009). From open source to open content: Organization, licensing and decision processes in open cultural production. *Decision Support Systems, 47*(3), 229–244.

Chesbrough, H. W. (2003). *Open innovation: The New Imperative for Creating and Profiting from Technology*. Boston: Harvard Business School Press.

Chi, L., Chan, W. K., Seow, G., & Tam, K. (2009). Transplanting social capital to the online world: Insights from two experimental studies. *Journal of Organizational Computing & Electronic Commerce, 19*(3), 214–236.

Chiu, C.-M., Hsu, M.-H., & Wang, E. T. G. (2006). Understanding knowledge sharing in virtual communities: An integration of social capital and social cognitive theories. *Decision Support Systems, 42*(3), 1872–1888.

Christensen, L. B. (2007). *Experimental methodology*. Boston: Allyn and Bacon.

Christiansen, J. K., & Varnes, C. J. (2009). Formal rules in product development: Sensemaking of structured approaches. *Journal of Product Innovation Management, 26*(5), 502–519.

Chua, A. Y. K. (2006). The rise and fall of a community of practice: A descriptive case study. *Knowledge & Process Management, 13*(2), 120–128.

Cindio, F. de, Gentile, O., Grew, P., & Redolfi, D. (2003). Community networks: Rules of behavior and social structure. *The Information Society, 19*(5), 395–406.

Clemens, W., & J. Strübing (Eds.) (2000). *Empirische Sozialforschung und gesellschaftliche Praxis*. Opladen: Leske + Budrich.

Cohen, S. G., & Bailey, D. E. (1997). What makes teams work: Group effectiveness research from the shop floor to the executive suite. *Journal of Management, 23*(3), 239–290.

Conceicao, P., Hamill, D., & Pinheiro, P. (2002). Innovative science and technology commercialization strategies at 3M: A case study. *Journal of Engineering & Technology Management, 19*(1), 25.

Cote, J. A., & Buckley (1987). Estimating trait, method, and error variance: Generalizing across 70 construct validation studies. *Journal of Marketing Research, 24*(3), 315–318.

Cothrel, J. P., & Williams, R. L. (1999). On-line communities: Helping them form and grow. *Strategy & Leadership, 28*(2), 17–21.

Cova, B., & Pace, S. (2006). Brand community of convenience products: New forms of customer empowerment – the case "my Nutella The Community". *European Journal of Marketing, 40*(9/10), 1087–1105.

Cromie, J., & Ewing, M. (2008). Squatting at the digital campfire. *International Journal of Market Research, 50*(5), 631–653.

Crossan, M. M., Lane, H. W., & White, R. E. (1999). An organizational learning framework: From intuition to institution. *Academy of Management Review, 24*(3), 522–537.

Dahlander, L., Frederiksen, L., & Rullani, F. (2008). Online communities and open innovation: Governance and symbolic value creation. *Industry & Innovation, 15*(2), 115–123.

Davies, B. (2005). Communities of practice: Legitimacy not choice. *Journal of Sociolinguistics, 9*(4), 557–581.

Deci, E. L., & Ryan, R. M. (2002). *Handbook of Self-determination Research.* Rochester, NY: University of Rochester Press.

DeLone, W. H., & McLean, E. R. (1992). Information systems success: The quest for the dependent variable. *Information Systems Research, 3*(1), 60–95.

Denison, D. R. (1984). Bringing corporate culture to the bottom line. *Organizational Dynamics, 13*(2), 4–22.

Denning, P. J. (1997). A new social contract for research. *Communications of the ACM, 40*(2), 132–134.

Desouza, K. C., Awazu, Y., Jha, S., Dombrowski, C., Papagari, S., Baloh, P., et al. (2008). Customer-driven innovation. *Research Technology Management, 51*(3), 35–44.

Dewhurst, F. W., & Cegarra Navarro, J. G. (2004). External communities of practice and relational capital. *Learning Organization, 11*(4/5), 322–331.

Dholakia, U. M., Bagozzi, R. P., & Pearo, L. K. (2004). A social influence model of consumer participation in network- and small-group-based virtual communities. *International Journal of Research in Marketing, 21*(3), 241–263.

Donaldson, S. I., & Grant-Vallone, E. J. (2002). Understanding self-report bias in organizational behavior research. *Journal of Business and Psychology, 17*(2), 245–260.

Donath, J. S. (2004). Identity and deception in the virtual community. In M. A. Smith & P. Kollock (Eds.), *Communities in Cyberspace* (pp. 29–59). London: Routledge.

Drath, W. H., & Palus, C. J. (1994). *Making Common Sense: Leadership as Meaning-making in a Community of Practice.* Greensboro, NC: Center for Creative Leadership.

Driver, M. (2002). Learning and leadership in organizations: Toward complementary communities of practice. *Management Learning, 33*(1), 99–126.

Dubé, L., Bourhis, A., & Jacob, R. (2005).The impact of structuring characteristics on the launching of virtual communities of practice. *Journal of Organizational Change Management, 18*(2), 145–166.

Dul, J., Bansemir, B., Ceylan, C., & Lundberg, H. (2011). Organizing creativity for innovation: Multidisciplinary perspectives, theories, and practices. In *European Academy of Managemen (EURAM).* Tallinn.

Durkheim, E. (1933). *The Division of Labor in Society.* Trans. G. Simpson.New York: The Free Press.

Ebner, W., Leimeister, J. M., & Krcmar, H. (2009). Community engineering for innovations: The ideas competition as a method to nurture a virtual community for innovations. *R&D Management, 39*(4), 342–356.

Eden, C., & Huxham, C. (1996). Action research for management research. *British Journal of Management, 7*(1), 75–86.

Edmondson, A. (1999). Psychological safety and learning behavior in work teams. *Administrative Science Quarterly, 44*(2), 350–383.

Eisenhardt, K. M. (1989). Building theories from case study research. *Academy of Management Review, 14*(4), 532–550.

Erler, H., Rieger, M., & Füller, J. (2009). Ideenmanagement und Innovation mit Social Networks. In A. Zerfaß & K. M. Möslein (Eds.), *Kommunikation als Erfolgsfaktor im Innovationsmanagement: Strategien im Zeitalter der Open Innovation* (pp. 391–401). Wiesbaden: Gabler.

Etzioni, A. (1968). *The Active Society: A Theory of Societal and Political Processes.* London: Collier-Macmillan.

Etzioni, A., & Etzioni, O. (1999). Face-to-face and computer-mediated communities: A comparative analysis. *The Information Society: An International Journal, 15*(4), 241–248.

Fang, Y., & Neufeld, D. (2009). Understanding sustained participation in open source software projects. *Journal of Management Information Systems, 25*(4), 9–50.

Farquhar, J., & Rowley, J. (2006). Relationships and online consumer communities. *Business Process Management Journal, 12*(2), 162–177.

Feng, J., Lazar, J., & Preece, J. (2004). Empathy and online interpersonal trust: A fragile relationship. *Behaviour & Information Technology, 23*(2), 97–106.

Fichter, K. (Ed.) (2006a). *Innovationen für eine nachhaltige Entwicklung. Innovation Communities: Die Rolle von Promotorennetzwerken bei Nachhaltigkeitsinnovationen.* Wiesbaden: Springer.

Fichter, K. (2006b). Innovation Communities: Die Rolle von Promotorennetzwerken bei Nachhaltigkeitsinnovationen. In K. Fichter (Ed.), *Innovationen für eine nachhaltige Entwicklung* (pp. 287–300). Wiesbaden: Springer.

Fichter, K. (2009). Innovation communities: The role of networks of promotors in Open Innovation. *R&D Management, 39*(4), 357–371.

Fiol, C. M., & O'Connor, E. J. (2002). When hot and cold collide in radical change processes: Lessons from community development. *Organization Science, 13*(5), 532–546.

Fischer, G. (2001a). Communities of interest: Learning through the interaction of multiple knowledge systems. In G. Fischer (Ed.), *24th Annual Information Systems Research Seminar In Scandinavia (IRIS'24).* Ulvik, Norway: Citeseer.

Fischer, G. (2001b). External and shareable artifacts as opportunities for social creativity in communities of interest. *Computational and Cognitive Models of Creative Design V (Heron Island'01)*, 67–89.

Fleming, L., & Waguespack, D. M. (2007). Brokerage, boundary spanning, and leadership in open innovation communities. *Organization Science, 18*(2), 165–180.

Fox, S. (2000). Communities of practice, Foucault and actor-network theory. *Journal of Management Studies, 37*(6), 853–867.

Franke, N., & Shah, S. (2003). How communities support innovative activities: An exploration of assistance and sharing among end-users. *Research Policy, 32*(1), 157–178.

Fredrickson, B. L. (1998). What good are positive emotions? *Review of General Psychology,* *2*(3), 300–319.

Frieß, M. R., Groh, G., Reinhardt, M., Forster, F., & Schlichter, J. (2012). Context-aware creativity support for corporate open innovation. *International Journal of Knowledge-Based Organizations, 2.*

Füller, J. (2006). Why consumers engage in virtual new product developments initiated by producers. *Advances in Consumer Research, 33*(1), 639–646.

Füller, J., Bartl, M., Ernst, H., & Muehlbacher, H. (2006). Community based innovation: How to integrate members of virtual communities into new product development. *Electronic Commerce Research, 6*(1), 57–73.

Füller, J., Matzler, K., & Hoppe, M. (2008). Brand community members as a source of innovation. *Journal of Product Innovation Management, 25*(6), 608–619.

Gammelgaard, J. (2010). Knowledge retrieval through virtual communities of practice. *Behaviour & Information Technology, 29*(4), 349–362.

Ganley, D., & Lampe, C. (2009). The ties that bind: Social network principles in online communities: Online Communities and Social Network. *Decision Support Systems, 47*(3), 266–274.

Garavan, T. N., Carbery, R., & Murphy, E. (2007). Managing intentionally created communities of practice for knowledge sourcing across organisational boundaries: Insights on the role of the CoP manager. *Learning Organization, 14*(1), 34–49.

Garrety, K., Robertson, P. L., & Badham, R. (2004). Integrating communities of practice in technology development projects. *International Journal of Project Management, 22*(5), 351–358.

Gassmann, O., & Zedtwitz, M. von (1999). New concepts and trends in international R&D organization. *Research Policy, 28*(2-3), 231–250.

Gassmann, O., & Enkel, E. (2006). Open Innovation. *Zeitschrift Führung und Organisation, 75*(3), 132–138.

Gerybadze, A. (2007). Gruppendynamik und Verstehen in Innovation Communities. In C. Herstatt (Ed.), *Management der frühen Innovationsphasen. Grundlagen – Methoden – neue Ansätze* (2nd ed., pp. 199–213). Wiesbaden: Gabler.

Giddens, A. (1979). *Central Problems in Social Theory: Action, Structure, and Contradiction in Social Analysis.:* Berkeley: University of California Press.

Giddens, A. (1984). *The Constitution of Society: Outline of the Theory of Structuration.* Berkeley: University of California Press.

Ginsburg, M., & Weisband, S. (2004). A framework for virtual community business success: The case of the internet chess club. In *Proceedings of the 37th Annual Hawaii International Conference on System Sciences* (pp. 196–205). New York: IEEE Computer Society Press.

Gioia, D. A., & Chittipeddi, K. (1991). Sensemaking in strategic change initiation. *Strategic Management Journal, 12*(6), 433–448.

Gioia, D. A., Thomas, J. B., Clark, S. M., & Chittipeddi, K. (1994). Symbolism and strategic change in academia: The dynamics of sensemaking and influence. *Organization Science, 5*(3), 363–383.

Gladstein, D. L. (1984). Groups in context: A model of task group effectiveness. *Administrative Science Quarterly, 29*(4), 499–517.

Gloor, P. A., Paasivaara, M., Schoder, D., & Willems, P. (2008). Finding collaborative innovation networks through correlating performance with social network structure. *International Journal of Production Research, 46*(5), 1357–1371.

Golden, B. R. (1992). The past is the past—or is it? The use of retrospective accounts as indicators of past strategy. *Academy of Management Journal, 35*(4), 848–860.

Goll, I., Sambharya, R. B., & Tucci, L. A. (2001). Top management team composition, corporate ideology, and firm performance. *Management International Review, 41*(2), 109–130.

Gongla, P., & Rizzuto, C. R. (2001). Evolving communities of practice: IBM Global Services experience. *IBM Systems Journal, 40*(4), 842–862.

Grace-Farfaglia, P., Dekkers, A., Sundararajan, B., Peters, L., & Park, S.-H. (2006). Multinational web uses and gratifications: Measuring the social impact of online community participation across national boundaries. *Electronic Commerce Research, 6*(1), 75–101.

Granovetter, M. S. (1973). The strength of weak ties. *American Journal of Sociology, 78*(6), 1360–1380.

Gray, J. A. (1981). A critique of Eysenck's theory of personality. In H. J. Eysenck (Ed.), *A Model for Personality* (pp. 246–276). Berlin: Springer.

Gray, J. A. (1990). Brain systems that mediate both emotion and cognition. *Cognition & Emotion, 4*(3), 269–288.

Grewal, R., Lilien, G. L., & Mallapragada, G. (2006). Location, location, location: How network embeddedness affects project success in open source systems. *Management Science, 52*(7), 1043–1056.

Gu, B., Konana, P., Rajagopalan, B., & Chen, H.-W.M. (2007). Competition among virtual communities and user valuation: The case of investing-related communities. *Information Systems Research, 18*(1), 68–85.

Gupta, A. K., & Wilemon, D. (1990). Improving R&D/Marketing relations: R&D's perspective. *R&D Management, 20*(4), 277–290.

Gustafsson, A., Johnson, M. D., & Roos, I. (2005). The effects of customer satisfaction, relationship commitment dimensions, and triggers on customer retention. *Journal of Marketing, 69*(3), 210–218.

Hall, H., & Graham, D. (2004). Creation and recreation: Motivating collaboration to generate knowledge capital in online communities. *International Journal of Information Management, 24*(3), 235–246.

Haller, J. B., Adamczyk, S., Bansemir, B., Bullinger, A. C., & Möslein, K. M. (2011). Tell me how good I am – An empirical investigation of the impact of peer feedback in IT-based innovation contests. In *International Product Development Management Conference (IPDMC). Innovation through design.* Delft.

Handley, K., Sturdy, A., Fincham, R., & Clark, T. (2006). Within and beyond communities of practice: Making sense of learning through participation, identity and practice. *Journal of Management Studies, 43*(3), 641–653.

Hardy, Q. (2005). Google thinks small. *Forbes, 176*(10), 198–202.

Harhoff, D., Henkel, J., & Hippel, E. von (2003). Profiting from voluntary information spillovers: How users benefit by freely revealing their innovations. *Research Policy, 32*(10), 1753–1769.

Hars, A., & Ou, S. (2002). Working for free? Motivations for participating in open-source projects. *International Journal of Electronic Commerce, 6*(3), 25–34.

Harzing, A.-W. (2011). *Harzing's publish or perish*. Melbourne: Tarma Software Research Ltd.

Hemetsberger, A., & Reinhardt, C. (2006). Learning and knowledge-building in open source communities: A social-experiential approach. *Management Learning, 37*(2), 187–214.

Hennig-Thurau, T. G. K. P., Walsh, G., & Gremler, D. D. (2004). Electronic world-of-mouth via consumer-opinion platforms: What motivates consumers to articulate themselves on the internet? *Journal of Interactive Marketing, 18*(1), 38–52.

Hertel, G., Niedner, S., & Herrmann, S. (2003). Motivation of software developers in open source projects: An internet-based survey of contributors to the Linux kernel. *Research Policy, 32*(7), 1159–1177.

Hevner, A. R., March, S. T., Park, J., & Ram, S. (2004). Design science in information systems research. *MIS Quarterly*, 75–105.

Hildreth, P., Kimble, C., & Wright, P. (1998). *Computer Mediated Communications and Communities of Practice*. In Proceedings of Ethicomp'98. Rotterdam, The Netherlands. 275–286.

Hill, R. C., & Levenhagen, M. (1995). Metaphors and mental models: Sensemaking and sensegiving in innovative and entrepreneurial activities. *Journal of Management, 21*(6), 1057.

Hill, S. A., & Birkinshaw, J. M. (2010). Idea sets: Conceptualizing and measuring a new unit of analysis in entrepreneurship research. *Organizational research methods, 13*(1), 85–113.

Hippel, E. v. (1994). "Sticky information" and the locus of problem solving: Implications for innovation. *Management Science, 40*(4), 429–439.

Hippel, E. v. (2005). *Democratizing Innovation*. Cambridge, MA: MIT Press.

Hippel, E. v., & Krogh, G. von (2003). Open source software and the "private-collective" innovation model: Issues for organization science. *Organization Science, 14*(2), 209–223.

Hislop, D. (2003). The complex relations between communities of practice and the implementation of technological innovations. *International Journal of Innovation Management, 7*(7), 163–188.

Hsu, M.-H., Ju, T. L., Yen, C.-H., & Chang, C.-M. (2007). Knowledge sharing behavior in virtual communities: The relationship between trust, self-efficacy, and outcome expectations. *International Journal of Human-Computer Studies, 65*(2), 153–169.

Hummel, J., & Lechner, U. (2002). Social profiles of virtual communities. *Proceedings of the 35th Hawaii International Conference on System Sciences*. New York: IEEE Computer Society Press.

Hung, K. H., & Yiyan Li, S. (2007). The influence of eWOM on virtual consumer communities: Social capital, consumer learning, and behavioral outcomes. *Journal of Advertising Research, 47*(4), 485–495.

Hussler, C., & Rondé, P. (2007). The impact of cognitive communities on the diffusion of academic knowledge: Evidence from the networks of inventors of a French university. *Research Policy, 36*(2), 288–302.

Ilgen, D. R., Hollenbeck, J. R., Johnson, M., & Jundt, D. (2005). Teams in organizations: From input-process-output models to IMOI models. *Annual Review of Psychology, 56*(1), 517–543.

Ilies, R., & Judge, T. A. (2005). Goal regulation across time: The effects of feedback and affect. *Journal of Applied Psychology, 90*(3), 453–467.

Iriberri, A., & Leroy, G. (2009). A life-cycle perspective on online community success. *ACM Computing Surveys, 41*(2), 1–29.

Isen, A. M. (2002). Missing in action in the AIM: Positive affect's facilitation of cognitive flexibility, innovation, and problem solving. *Psychological Inquiry, 13*(1), 57–65.

Isen, A. M. (2001). An influence of positive affect on decision making in complex situations: Theoretical issues with practical implications. *Journal of Consumer Psychology (Lawrence Erlbaum Associates), 11*(2), 75–85.

Isen, A. M., Niedenthal, P. M., & Cantor, N. (1992). An influence of positive affect on social categorization. *Motivation and Emotion, 16*(1), 65–78.

Ishii, K., & Ogasahara, M. (2007). Links between real and virtual networks: A comparative study of online communities in Japan and Korea. *CyberPsychology & Behavior, 10*(2), 252–257.

Jäkälä, M., & Pekkola, S. (2007). From technology engineering to social engineering: 15 years of research on virtual worlds. *SIGMIS Database, 38*(4), 11–16.

Jakubik, M. (2008). Experiencing collaborative knowledge creation processes. *Learning Organization, 15*(1), 5–25.

Jang, H., Olfman, L., Ko, I., Koh, J., & Kim, K. (2008). The influence of on-line brand community characteristics on community commitment and brand loyalty. *International Journal of Electronic Commerce, 12*(3), 57–80.

Jarzabkowski, P. (2008). Shaping strategy as a structuration process. *Academy of Management Journal, 51*(4), 621–650.

Jian, G., & Jeffres, L. W. (2006). Understanding employees' willingness to contribute to shared electronic databases: A three-dimensional framework. *Communication Research, 33*(4), 242–261.

Jones, M. R., & Karsten, H. (2008). Gidden's structuration theory and information systems research. *MIS Quarterly, 32*(1), 127–157.

Jones, Q., & Rafaeli, S. (2000). What do virtual "tells" tell? Placing cybersociety research into a hierarchy of social explanation. In *Proceedings of the 33rd Hawaii International Conference on System Sciences*. New York: IEEE Computer Society Press.

Koh, J., & Young-Gul Kim (2003). Sense of virtual community: A conceptual framework and empirical validation. *International Journal of Electronic Commerce, 8*(2), 75–93.

Kahai, S. S., Carroll, E., & Jestice, R. (2007). Team collaboration in virtual worlds. *SIGMIS Database, 38*(4), 61–68.

Kannan, P. K., Chang, A.-M., & Whinston, A. B. (2000). Electronic communities in e-business: Their role and issues. *Information Systems Frontiers, 1*(4), 415–426.

Kasper, H., Mühlbacher, J., & Müller, B. (2008). Intra-organizational knowledge sharing in MNCs depending on the degree of decentralization and communities of practice. *Journal of Global Business & Technology, 4*(1), 59–68.

Katz, R., & Allen, T. J. (1982). Investigating the Not Invented Here (NIH) syndrome: A look at the performance, tenure, and communication patterns of 50 R&D project groups. *R&D Management, 12*(1), 7–20.

Kavanagh, D., & Kelly, S. (2002). Sensemaking, safety, and situated communities in (con)temporary networks. *Journal of Business Research, 55*(7), 583–594.

Ke, W., & Zhang, P. (2009). Motivations in open source software communities: The mediating role of effort intensity and goal commitment. *International Journal of Electronic Commerce, 13*(4), 39–66.

Kerr, S., & Jermier, J. M. (1978). Substitutes for leadership: Their meaning and measurement. *Organizational Behavior & Human Performance, 22*(3), 375–403.

Kim, J., Song, J., & Jones, D. R. (2011). The cognitive selection framework for knowledge acquisition strategies in virtual communities. *International Journal of Information Management, 31*(2), 111–120.

Kimble, C., & Bourdon, I. (2008). Some success factors for the communal management of knowledge. *International Journal of Information Management, 28*(6), 461–467.

Kinnear, T. C. & Taylor, J. R. (1983). *Marketing Research: An Applied Approach.* New York: McGraw-Hill.

Kirschten, U. (2006). Nachhaltige Innovationsnetzwerke in Theorie und Praxis: Ausgewählte Forschungsergebnisse. In R. Pfriem et al. (Eds.), *Innovationen für eine nachhaltige Entwicklung* (pp. 269–286). Wiesbaden: Gabler.

Kling, R., & Courtright, C. (2003). Group behavior and learning in electronic forums: A sociotechnical approach. *Information Society, 19*(3), 221–235.

Knoben, J., & Oerlemans, L. (2006). Proximity and inter-organizational collaboration: A literature review. *International Journal of Management Reviews, 8*(2), 71–89.

Kodama, M. (2001a). Creating new business through strategic community management: Case study of a multimedia business. *International Journal of Human Resource Management, 12*(6), 1062–1084.

Kodama, M. (2001b). Innovation through strategic community management: The case of NTT DoCoMo and the mobile internet revolution. *Creativity & Innovation Management, 10*(2), 75–87.

Kodama, M. (2005a). How two Japanese high-tech companies achieved rapid innovation via strategic community networks. *Strategy & Leadership, 33*(6), 39–47.

Kodama, M. (2005b). New knowledge creation through leadership-based strategic community—a case of new product development in IT and multimedia business fields. *Technovation, 25*(8), 895–908.

Kodama, M. (2006). Strategic community: Foundation of knowledge creation. *Research Technology Management, 49*(5), 49–58.

Koh, J., & Kim, Y.-G. (2003). Sense of virtual community: A conceptual framework and empirical validation. *International Journal of Electronic Commerce, 8*(2), 75–93.

Komito, L. (1998). The Net as a foraging society: Flexible communities. *The Information Society, 14*(2), 97–106.

Krieger, B. L., & Müller, P. S. (2003). Making internet communities work: Reflections on an unusual business model. *SIGMIS Database, 34*(2), 50–59.

Krippendorff, K. (1980). *Content Analysis: An Introduction to its Methodology.* Beverly Hills: Sage.

Krogh, G. von, Spaeth, S., & Lakhani, K. R. (2003). Community, joining, and specialization in open source software innovation: a case study: Open source software development. *Research Policy, 32*(7), 1217–1241.

Krumsvik, R. (2005). ICT and community of practice. *Scandinavian Journal of Educational Research, 49*(1), 27–50.

Kuo, Y.-F.(2003). A study on service quality of virtual community websites. *Total Quality Management & Business Excellence, 14*(4), 461–474.

Lakhani, K. R., Jeppesen, L. B., Lohse, P. A., & Panetta, J. A. (2007). *The Value of Openness in Scientific Problem Solving.* Citeseer.

Lakhani, K. R., & Hippel, E. v. (2003). How open source software works: "Free" user-to-user assistance. *Research Policy, 32*(6), 923–943.

Lamnek, S. (2000). Sozialforschung in Theorie und Praxis. Zum Verhältnis von qualitativer und quantitativer Forschung. In W. Clemens & J. Strübing (Eds.), *Empirische Sozialforschung und gesellschaftliche Praxis* (pp. 23–46). Opladen: Leske + Budrich.

Langley, A. (1999). Strategies for theorizing from process data. *Academy of Management Review, 24*(4), 691–710.

Lank, E., Randell-Khan, J., Rosenbaum, S., & Tate, O. (2008). Herding cats: Choosing a governance structure for your communities of practice. *Journal of Change Management, 8*(2), 101–109.

Lashinky, A. (2006). Chaos by design. *Fortune, 154*(7), 86–98.

Lave, J., & Wenger, E. (1991). *Situated Learning: Legitimate Peripheral Participation*. Cambridge: Cambridge University Press.

Lawrence, P. R., & Lorsch, J. W. (1967). Differentiation and integration in complex organizations. *Administrative Science Quarterly, 12*(1), 1–47.

Lea, B.-R., Yu, W.-B., Maguluru, N., & Nichols, M. (2006). Enhancing business networks using social network based virtual communities. *Industrial Management & Data Systems, 106*(1), 121–138.

Lee, H., & Choi, B. (2003). Knowledge management enablers, processes, and organizational performance: An integrative view and empirical examination. *Journal of Management Information Systems, 20*(1), 179–228.

Lee, S. H., & Williams, C. (2007). Dispersed entrepreneurship within multinational corporations: A community perspective. *Journal of World Business, 42*(4), 505–519.

Leimeister, J. M., Ebner, W., & Krcmar, H. (2005). Design, implementation, and evaluation of trust-supporting components in virtual communities for patients. *Journal of Management Information Systems, 21*(4), 101–135.

Leimeister, J. M., & Krcmar, H. (2004). Revisiting the virtual community business model. *Proceedings of the Tenth Americas Conference on Information Systems*, New York.

Leimeister, J. M., Schweizer, K., Leimeister, S., & Krcmar, H. (2008). Do virtual communities matter for the social support of patients? Antecedents and effects of virtual relationships in online communities. *Information Technology & People, 21*(4), 350–374.

Leimeister, J. M., Sidiras, P., & Krcmar, H. (2006). Exploring success factors of virtual communities: The perspectives of members and operators. *Journal of Organizational Computing & Electronic Commerce, 16*(3/4), 279–300.

Lerner, J., & Tirole, J. (2002). Some simple economics of open source. *Journal of Industrial Economics, 50*(2), 197–234.

Lesser, E. L., & Storck, J. (2001). Communities of practice and organizational performance. *IBM Systems Journal, 40*(4), 831–841.

Lin, H. F. (2007). Knowledge sharing and firm innovation capability: An empirical study. *International Journal of Manpower, 28*(3/4), 315–332.

Lin, H.-F. (2008). Antecedents of virtual community satisfaction and loyalty: An empirical test of competing theories. *CyberPsychology & Behavior, 11*(2).

Lin, H.-F., & Lee, G.-G. (2006). Determinants of success for online communities: An empirical study. *Behaviour & Information Technology, 25*(6), 279–488.

Lin, Y.-R., Chi, Y., Zhu, S., Sundaram, H., & Tseng, B. L. (2009). Analyzing communities and their evolutions in dynamic social networks. *ACM Transactions on Knowledge Discovery from Data, 3*(2), 1–31.

Lindkvist, L. (2005). Knowledge communities and knowledge collectivities: A typology of knowledge work in groups. *Journal of Management Studies, 42*(6), 1189–1210.

Linehan, C., & McCarthy, J. (2001). Reviewing the 'community of practice' metaphor: An analysis of control relations in a primary school classroom. *Mind, Culture, and Activity, 8*(2), 129–147.

Ludford, P. J., Cosley, D., Frankowski, D., & Terveen, L. (2004). Think different: Increasing online community participation using uniqueness and group dissimilarity. *chi 2004*.

Lueg, C. (Ed.). 2000. *Where is the action in virtual communities of practice: Proceedings of the Workshop Communication and Cooperation in Knowledge Communities at the D-CSCW*. Citeseer.

Luhmann, N. (1979). *Trust and Power*. Chichester: Wiley.

Lüscher, L. S., & Lewis, M. W. (2008). Organizational change and managerial sensemaking: Working through paradox. *Academy of Management Journal, 51*(2), 221–240.

Lynn, L. H., Aram, J. D., & Mohan Reddy, N. (1997). Technology communities and innovation communities. *Journal of Engineering and Technology Management, 14*(2), 129–145.

Lynn, L. H., Mohan Reddy, N., & Aram, J. D. (1996). Linking technology and institutions: The innovation community framework. *Research Policy, 25*(1), 91–106.

Ma, M., & Agarwal, R. (2007). Through a glass darkly: Information technology design, identity verification, and knowledge contribution in online communities. *Information Systems Research, 18*(1), 42–67.

MacDonald, R. J. (2008). Professional development for information communication technology integration: Identifying and supporting a community of practice through design-based research. *Journal of Research on Technology in Education, 40*(4), 429–445.

Maitlis, S. (2005). The social processes of organizational sensemaking. *Academy of Management Journal, 48*(1), 21–49.

Maitlis, S., & Lawrence, T. B. (2007). Triggers and enablers of sensegiving in organizations. *Academy of Management Journal, 50*(1), 57–84.

Maloney-Krichmar, D., & Preece, J. (2005). A multilevel analysis of sociability, usability, and community dynamics in an online health community. *ACM Transactions on Computer-Human Interaction, 12*(2), 201–232.

Marett, K., & Joshi, K. D. (2009). The decision to share information and rumors: Examining the role of motivation in an online discussion forum. *Communications of the Association for Information Systems, 24*(1), 44–69.

Marks, M. A., Mathieu, J. E., & Zaccaro, S. J. (2001). A temporally based framework and taxonomy of team processes. *Academy of Management Review, 26*(3), 356–376.

Marsden, P., & Campbell, K. (1984). Measuring tie strength. *Social Forces, 63*(2), 482–501.

Martin, L. L., Ward, D. W., Achee, J. W., & Wyer, R. S. (1993). Mood as input: People have to interpret the motivational implications of their moods. *Journal of Personality and Social Psychology, 64*(3), 317–326.

Matei, S. (2004). The impact of state-level social capital on the emergence of virtual communities. *Journal of Broadcasting & Electronic Media, 48*(1), 23–40.

Mathieu, J., Maynard, M. T., Rapp, T., & Gilson, L. (2008). Team effectiveness 1997–2007: A review of recent advancements and a glimpse into the future. *Journal of Management, 34*(3), 410–476.

Mathwick, C., Wiertz, C., & Ruyter, K. de (2008). Social capital production in a virtual P3 community. *Journal of Consumer Research, 34*(6), 832–849.

Maxwell, S. E., & Delaney, H. D. (2004). *Designing experiments and analyzing data: A model comparison perspective.* Mahwah, NJ: Erlbaum.

Mayring, P. (2002). *Einführung in die qualitative Sozialforschung: Eine Anleitung zu qualitativem denken:* Beltz.

McDermott, R. (1999a). Nurturing three-dimensional communities of practice. *Knowledge Management Review,* 26–29.

McDermott, R. (1999b). Why information technology inspired but cannot deliver knowledge management. *California Management Review, 41*(4), 103–117.

McGregor, J. (2006). The world's most innovative companies. *Business Week, 24*(04).

McMillan, D. W., & Chavis, D. M. (1986). Sense of community: A definition and theory. *Journal of Community Psychology, 14*(1), 6–23.

Mican, D., Tomai, N., & Coros, R. I. (2009). Web content management systems, a collaborative environment in the information society. *Informatica Economica, 13*(2), 20–31.

Miles, M. B., & A. M. Huberman (1994). *Qualitative Data Analysis: An Expanded Sourcebook.* Thousand Oaks, CA: Sage.

Millen, D. R., Fontaine, M. A., & Muller, M. J. (2002). Understanding the benefit and costs of communities of practice. *Communications of the ACM, 45*(4), 69–73.

Money, R. B., Gilly, M. C., & Graham, J. L. (1998). Explorations of national culture and word-of-mouth referral behavior in the purchase of industrial services in the United States and Japan. *The Journal of Marketing, 62*(4), 76–87.

MØrk, B. E., Aanestad, M., Hanseth, O., & Grisot, M. (2008). Conflicting epistemic cultures and obstacles for learning across communities of practice. *Knowledge & Process Management, 15*(1), 12–23.

Möslein, K. M., & Bansemir, B. (2008). Open Innovation als Innovationsstrategie. In I. Gatermann (Ed.), *Technologie und Dienstleistung.Innovationen in Forschung, Wissenschaft und Unternehmen* (pp. 301–310). Frankfurt am Main: Campus-Verl.

Möslein, K. M., & Bansemir, B. (2010). Strategic open innovation: Basics, actors, tools and tensions. In M. Hülsmann & N. Pfeffermann (Eds.), *Strategies and Communications for Innovations.* Berlin: Springer.

Möslein, K. M., Bansemir, B., & Haller, J. B. (2011). IT-unterstützte Methoden der Open Innovation. In M. Amberg & M. Lang (Eds.), *Wertorientierte IT.*

Möslein, K. M., & Neyer, A.-K. (2009). Open Innovation: Grundlagen, Herausforderungen, Spannungsfelder. In A. Zerfaß & K. M. Möslein (Eds.), *Kommunikation als Erfolgsfaktor im Innovationsmanagement: Strategien im Zeitalter der Open Innovation* (pp. 85–103). Wiesbaden: Gabler.

Muller, P. (2006). Reputation, trust and the dynamics of leadership in communities of practice. *Journal of Management & Governance, 10*(4), 381–400.

Mutch, A. (2003). Communities of practice and habitus: A critique. *Organization Studies, 24*(3), 383–401.

Nambisan, S. (2002). Designing virtual customer environments for new product development: Toward a theory. *Academy of Management Review, 27*(3), 392–413.

Neyer, A.-K., Bullinger, A. C., & Möslein, K. M. (2009). Integrating inside and outside innovators: A sociotechnical systems perspective. *R&D Management, 39*(4), 410–419.

Nonaka, I. (1994). A dynamic theory of organizational knowledge creation. *Organization Science, 5*(1), 14–37.

Nonaka, I., & Takeuchi, H. (1995). *The knowledge Creating Company: How Japanese Companies Create the Dynamics of Innovation.* New York: Oxford University Press.

Nonaka, I., & Toyama, R. (2003). The knowledge-creating theory revisited: Knowledge creation as a synthesizing process. *Knowledge Management Research & Practice, 1*(1), 2–10.

Nonnecke, B., Andrews, D., & Preece, J. (2006). Non-public and public online community participation: Needs, attitudes and behavior. *Electronic Commerce Research, 6*(1), 7–20.

Nonnecke, B., & Preece, J. (2001). Why lurkers lurk. *Americas Conference on Information Systems (AMCIS) 2001*, 1–10.

Nooteboom, B. (2000). Learning by interaction: Absorptive capacity, cognitive distance and governance. *Journal of Management and Governance, 4*(1), 69–92.

Novak, J. (2007). Multiperspektivische Wissensvisualisierung für Wissensaustausch in heterogenen Community-Netzwerken. *Institut für Informatik, FG Informationsmanagement*, 1–18.

Novicevic, M. M., Harvey, M. G., Buckley, M. R., Wren, D., & Pena, L. (2007). Communities of creative practice: Follett's seminal conceptualization. *International Journal of Public Administration, 30*(4), 367–385.

Oh, W., & Jeon, S. (2007). Membership herding and network stability in the open source community: The Ising perspective. *Management Science, 53*(7), 1086–1101.

O'Mahony, S. (2007). The governance of open source initiatives: What does it mean to be community managed? *Journal of Management & Governance, 11*(2), 139–150.

O'Mahony, S., & Ferraro, F. (2007). The emergence of governance in an open source community. *Academy of Management Journal, 50*(5), 1079–1106.

Orlikowski, W. J. (1996). Improvising organizational transformation over time: A situated change perspective. *Information Systems Research, 7*(1), 63–92.

Orr, J. E. (1986). Narratives at work: Story telling as cooperative diagnostic activity. In *Proceedings of the 1986 ACM conference on Computer-supported cooperative work* (pp. 62–72).

Osterloh, M., & Frey, B. S. (2000). Motivation, knowledge transfer, and organizational forms. *Organization Science, 11*(5), 538–550.

Østerlund, C., & Carlile, P. (2005). Relations in practice: Sorting through practice theories on knowledge sharing in complex organizations. *Information Society, 21*(2), 91–107.

Otto, P., & Simon, M. (2008). Dynamic perspectives on social characteristics and sustainability in online community networks. *System Dynamics Review (Wiley), 24*(3), 321–347.

Palincsar, A. S., Magnusson, S. J., Marano, N., Ford, D., & Brown, N. (1998). Designing a community of practice: principles and practices of the GIsML community. *Teaching and Teacher Education, 14*(1), 5–19.

Pan, S. L., & Leidner, D. E. (2003). Bridging communities of practice with information technology in pursuit of global knowledge sharing. *The Journal of Strategic Information Systems, 12*(1), 71–88.

Panzarasa, P., Opsahl, T., & Carley, K. M. (2009). Patterns and dynamics of users' behavior and interaction: Network analysis of an online community. *Journal of the American Society for Information Science & Technology, 60*(5), 911–932.

Patriotta, G. (2003). Sensemaking on the shop floor: Narratives of knowledge in organizations. *Journal of Management Studies, 40*(2), 349–375.

Peltonen, T., & Lämsä, T. (2004). 'Communities of practice' and the social process of knowledge creation: Towards a new vocabulary for making sense of organizational learning. *Problems & Perspectives in Management*, (4), 249–262.

Pemberton, J., Mavin, S., & Stalker, B. (2007). Scratching beneath the surface of communities of (mal)practice. *Learning Organization, 14*(1), 62–73.

Pfriem, R., Antes, R. Fichter, K., Müller, M., Paech, N., Seuring, S., & Siebenhüner, B. (Eds.) (2006). *Innovationen für eine nachhaltige Entwicklung*. Wiesbaden: Gabler.

Pisano, G. P., & Verganti, R. (2008). Which kind of collaboration is right for you? *Harvard Business Review, 86*(12), 79–86.

Plaskoff, J. (2003). Intersubjectivity and community building: Learning to learn organizationally. In M. Easterby-Smith (Ed.), *The Blackwell Handbook of Organizational Learning and Knowledge Management* (pp. 161–184). Malden, MA: Blackwell.

Podsakoff, P. M., MacKenzie, S. B., Lee, J. Y., & Podsakoff, N. P. (2003). Common method biases in behavioral research: A critical review of the literature and recommended remedies. *Journal of Applied Psychology, 88*(5), 879–904.

Polanyi, M. (1966). *The Tacit Dimension*. Gloucester, MA: Smith.

Porter, C. E., & Donthu, N. (2008). Cultivating trust and harvesting value in virtual communities. *Management Science, 54*(1), 113–128.

Preece, J. (2001). Sociability and usability in online communities: Determining and measuring success. *Behaviour & Information Technology, 20*(5), 347–356.

Probst, G., & Borzillo, S. (2008). Why communities of practice succeed and why they fail. *European Management Journal, 26*(5), 335–347.

Quigley, N. R., Tesluk, P. E., Locke, E. A., & Bartol, K. M. (2007). A multilevel investigation of the motivational mechanisms underlying knowledge sharing and performance. *Organization Science, 18*(1), 71–88.

Raymond, E. (1999). The cathedral and the bazaar. *Knowledge, Technology & Policy, 12*(3), 23–49.

Reason, P. (Ed.) (1988). *Human Inquiry in Action: Developments in New Paradigm Research.* London: Sage.

Reinhardt, M., Wiener, M., Frieß, M. R., Groh, G., & Amberg, M. (2012). Social software support for collaborative innovation development within organizations. *International Journal of Knowledge-Based Organizations, 2*.

Reinmoeller, P., & Chong, L.-C. (2002). Managing the knowledge-creating context: A strategic time approach. *Creativity & Innovation Management, 11*(3), 165–174.

Richter, A., Mörl, S., & Koch, M. (2011). Anwendungsszenarien als Werkzeug zur (V)Ermittlung des Nutzens von Corporate Social Software. In A. Bernstein & G. Schwabe (Eds.), *Proceedings of the 10th International Conference on Wirtschaftsinformatik. WI 2.011* (pp. 1104–1122). Zurich: Lulu.

Ridings, C., Gefen, D., & Arinze, B. (2006). Psychological barriers: Lurker and poster motivation and behavior in online communities. *Communications of AIS, 2006*(18), 329–354.

Ridings, C. M., Gefen, D., & Arinze, B. (2002). Some antecedents and effects of trust in virtual communities. *The Journal of Strategic Information Systems, 11*(3-4), 271–295.

Roberts, J. A., Hann, I.-H., & Slaughter, S. A. (2006). Understanding the motivations, participation, and performance of open source software developers: A longitudinal study of the Apache projects. *Management Science, 52*(7), 984–999.

Roberts, J. (2006). Limits to communities of practice. *Journal of Management Studies, 43*(3), 623–639.

Rollinson, D. (2008). *Organisational Behaviour and Analysis: An Integrated Approach.* Harlow: Financial Times & Prentice Hall.

Romm, C., & Pliskin, N. (1997). Virtual communities and society: Toward an integrative three phase model. *International Journal of Information Management, 17*(4), 261–270.

Roschek, J. (2009). Web 2.0 als Innovationsplattform. In A. Zerfaß & K. M. Möslein (Eds.), *Kommunikation als Erfolgsfaktor im Innovationsmanagement: Strategien im Zeitalter der Open Innovation* (pp. 379–389). Wiesbaden: Gabler.

Rosenbaum, H., & Shachaf, P. (2010). A structuration approach to online communities of practice: The case of Q&A communities. *Journal of the American Society for Information Science & Technology, 61*(9), 1933–1944.

Rothaermela, F. T., & Sugiyamab, S. (2001). Virtual internet communities and commercial success: Individual and community-level theory grounded in the atypical case of TimeZone.com. *Journal of Management, 27*(3), 297–312.

Rothwell, R. (1994). Towards the fifth-generation innovation process. *International Marketing Review, 11*(1), 7–31.

Ryan, G. W., & Bernard, H. R. (2000). Data management and analysis methods. *Handbook of Qualitative Research, 2,* 769–802.

Sangwan, S. (2005). Virtual community success: A uses and gratifications perspective. *Proceedings of the 38th Hawaii International Conference on System Sciences.* New York: IEEE Computer Society Press.

Sarris, V. (1990). *Methodologische Grundlagen der Experimentalpsychologie.* München: E. Reinhardt.

Sawhney, M., & Prandelli, E. (2000). Communities of creation: Managing distributed innovation in turbulent markets. *California Management Review, 42*(4), 24–54.

Schoberth, T., Heinzl, A., & Preece, J. (2006). Exploring communication activities in online communities: A longitudinal analysis in the financial services industry. *Journal of Organizational Computing & Electronic Commerce, 16*(3/4), 247–265.

Schwarz, S., & Bodendorf, F. (2012). Attributive idea evaluation – A new idea evaluation method for corporate open innovation communities. *International Journal of Knowledge-Based Organizations, 2.*

Schwarzer, R., & Jerusalem, M. (1995). General perceived self-efficacy. In J. Weinmann, S. Wright, & M. Johnston (Eds.), *Measures in Health Psychology: A User's Portfolio. Causal and Control Beliefs* (pp. 35–37). Windsor: NFER-Nelson.

Schwen, T. M., & Hara, N. (2003). Community of practice: A metaphor for online design? *Information Society, 19*(3), 257–270.

Shah, S. K. (2006). Motivation, governance, and the viability of hybrid forms in open source software development. *Management Science, 52*(7), 1000–1014.

Siebert, J. (2006). *Führungssysteme zwischen Stabilität und Wandel: Ein systematischer Ansatz zum Management der Führung.* Wiesbaden: Gabler.

Silva, L., Goel, L., & Mousavidin, E. (2009). Exploring the dynamics of blog communities: the case of MetaFilter. *Information Systems Journal, 19*(1), 55–81.

Sloman, M., & Reynolds, J. (2003). Developing the e-learning community. *Human Resource Development International, 6*(2), 259–272.

Smith, M. (1998). The development of an innovation culture. *Management Accounting London, 76*, 22–25.

Soekijad, M., Huis in't Veld, M. A. A., & Enserink, B. (2004). Learning and knowledge processes in inter-organizational communities of practice. *Knowledge & Process Management, 11*(1), 3–12.

Soto, J. P., Vizcaíno, A., Portillo-Rodríguez, J., & Piattini, M. (2007). Applying trust, reputation and intuition aspects to support virtual communities of practice. In *Lecture Notes in Computer Science*(pp. 353–360). Berlin, Heidelberg: Springer.

Stanley, J. C., & Campbell, D. T. (1966). *Experimental and Quasi-experimental Designs for Research.* Newbury Park: Sage.

Stanoevska-Slabeva, K. (2002). Toward a community-oriented design of internet platforms. *International Journal of Electronic Commerce, 6*(3), 71–95.

Star, S. L. (1989). The structure of ill-structured solutions: Heterogeneous problem solving, boundary objects and heterogeneous distributed problem solving. In M. Huhns & L. Gasser (Eds.), *Distributed Artificial Intelligence* (pp. 37–54). London: Pitman Publishing.

Star, S. L., & Griesemer, J. R. (1989). Institutional ecology, 'translations' and boundary objects: Amateurs and professionals in Berkeley's museum of vertebrate zoology, 1907-39. *Social Studies of Science, 19*(3), 387–420.

Starbuck, W. H., & Milliken, F. J. (1988). Executives' perceptual filters: What they perceive and how they make sense. *The Executive Effect: Concepts and Methods for Studying Top Managers*, 35–65.

Swan, J. A., Scarbrough, H., & Robertson, M. (2002). The construction of 'communities of practice' in the management of innovation. *Management Learning, 33*(4), 477–496.

Szmigin, I., Canning, L., & Reppel, A. E. (2005). Online community: Enhancing the relationship marketing concept through customer bonding. *International Journal of Service Industry Management, 16*(5), 480–496.

Szulanski, G. (2000). The process of knowledge transfer: A diachronic analysis of stickiness. *Organizational Behavior & Human Decision Processes, 82*(1), 9–27.

Szulanski, G., & Cappetta, R. (2003). Stickiness: Conceptualizing, measuring, and predicting difficulties in the transfer of knowledge within organizations. In M. Easterby-Smith (Ed.), *The Blackwell Handbook of Organizational Learning and Knowledge Management* (pp. 513–533). Malden, MA: Blackwell.

Szulanski, G., Cappetta, R., & Jensen, R. J. (2004). When and how trustworthiness matters: Knowledge transfer and the moderating effect of causal ambiguity. *Organization Science, 15*(5), 600–613.

Talukder, M., & Yeow, P. H. (2007). A comparative study of virtual communities in bangladesh and the USA. *Journal of Computer Information Systems, 47*(4), 82–90.

Tarmizi, H., & Vreede, G. J. de (Eds.). (2005). *A facilitation task taxonomy for communities of practice.* Omaha, NE.

Tedjamulia, S. J. J., Dean, D. L., Olsen, D. R., & Albrecht, C. C. (2005). Motivating content contributions to online communities: Toward a more comprehensive theory.

Proceedings of the 38th Hawaii International Conference on System Sciences. New York: IEEE Computer Society Press.

Thompson, M. (2005). Structural and epistemic parameters in communities of practice. *Organization Science, 16*(2), 151–164.

Torbert, W. R. (1976). *Creating a Community of Inquiry*. London: Wiley.

Tsichritzis, D. (1998). *The Dynamics of Innovation*. New York: Copernicus Books.

Tyler, J., Wilkinson, D., & Huberman, B. (2005). E-Mail as spectroscopy: Automated discovery of community structure within organizations. *Information Society, 21*(2), 133–141.

Valck, K. de, Langerak, F., Verhoef, P. C., & Verlegh, P. W. J. (2007). Satisfaction with virtual communities of interest: Effect on members' visit frequency. *British Journal of Management, 18*(3), 241–256.

Valck, K. de, van Bruggen, G. H., & Wierenga, B. (2009). Virtual communities: A marketing perspective: Online communities and social network. *Decision Support Systems, 47*(3), 185–203.

van Oost, E., Verhaegh, S., & Oudshoorn, N. (2009). From innovation community to community innovation: User-initiated innovation in wireless Leiden. *Science, Technology & Human Values, 34*(2), 182–205.

Vassileva, J., & Sun, L. (2007). Using community visualization to stimulate participation in online communities. *e-Service Journal, 6*(1), 3–39.

Velamuri, V. (2011). *Hybrid Value Creation*. Weisbaden: Gabler Verlag.

Venters, W., & Wood, B. (2007). Degenerative structures that inhibit the emergence of communities of practice: A case study of knowledge management in the British Council. *Information Systems Journal, 17*(4), 349–368.

Verburg, R. M., & Andriessen, J. H. E. (2006). The assessment of communities of practice. *Knowledge & Process Management, 13*(1), 13–25.

Wachter, R. M., Gupta, J. N. D., & Quaddus, M. A. (2000). IT takes a village: Virtual communities in support of education. *International Journal of Information Management, 20*(6), 473–489.

Wagenaar, S., & Hulsebosch, J. (2008). From 'a meeting' to 'a learning community'. *Group Facilitation: A Research & Applications Journal, 9*, 14–36.

Wamalwa, T. (2007). Internet technology and challenges of virtual communities. *International Journal of Business Research, 7*(4), 69–85.

Wang, J.-C., & Chen, C.-L.(2004). An automated tool for managing interactions in virtual communities – Using social network analysis approach. *Journal of Organizational Computing & Electronic Commerce, 14*(1), 1–26.

Wang, P., & Ramiller, N. C. (2009). Community learning in information technology innovation. *MIS Quarterly, 33*(4), 709–734.

Wang, Y., & Fesenmaier, D. R. (2004). Modeling participation in an online travel community.*Journal of Travel Research, 42*(3), 261–270.

Wasko, M. M., Teigland, R., & Faraj, S. (2009). The provision of online public goods: Examining social structure in an electronic network of practice. *Decision Support Systems, 47*(3), 254–265.

Watson, D., Wiese, D., Vaidya, J., & Tellegen, A. (1999). The two general activation systems of affect: Structural findings, evolutionary considerations, and psychobiological evidence. *Journal of Personality & Social Psychology, 76*(5), 820–838.

Weick, K. E. (1995). *Sensemaking in Organizations*. Thousand Oaks, CA: Sage.

Weick, K. E. (1988). Enacted sensemaking in crisis situations. *Journal of Management Studies, 25*(4), 305–317.

Weick, K. E. (1993). The collapse of sensemaking in organizations: The Mann Gulch disaster. *Administrative Science Quarterly, 38*(4), 628–652.

Weick, K. E. (2002). Puzzles in organizational learning: An exercise in disciplined imagination. *British Journal of Management, 13*, S7-S15.

Weick, K. E., Sutcliffe, K. M., & Obstfeld, D. (2005). Organizing and the process of sensemaking. *Organization Science, 16*(4), 409–421.

Wellman, B., Haase, A. Q., Witte, J., & Hampton, K. (2001). Does the internet increase, decrease, or supplement social capital? Social networks, participation, and community commitment. *American Behavioral Scientist, 45*(3), 436–455.

Wenger, E. (1998). *Communities of Practice: Learning, Meaning and Identity. Learning in Doing: Social, Cognitive, and Computational Perspectives.* Cambridge: Cambridge University Press.

Wenger, E. (2000). Communities of practice and social learning systems. *Organization, 7*(2), 225–246.

Wenger, E., McDermott, R., & Snyder, W. (2002). *Cultivating Communities of Practice: A Guide to Managing Knowledge.* Boston, MA: Harvard Business School Press.

Wenger, E. C., & Snyder, W. M. (2000). Communities of practice: The organizational frontier. *Harvard Business Review, 78*(1), 139–145.

Wilson, J. M., Goodman, P. S., & Cronin, M. A. (2007). Group learning. *Academy of Management Review, 32*(4), 1041–1059.

Witte, E. (1972). *Das Informationsverhalten in Entscheidungsprozessen.* Mohr.

Witzeman, S., Slowinski, G., Dirkx, R., Gollob, L., Tao, J., Ward, S., et al. (2006). Harnessing external technology for innovation. *Research Technology Management, 49*(3), 19–27.

Wood, R., & Bandura, A. (1989).Social cognitive theory of organizational management.*Academy of Management Review, 14*(3), 361–384.

Woodland, D. E., Szul, L. F., & Moore, W. A. (2007). Virtual learning communities. *Business Education Digest*, (16), 70–80.

Yin, R. K. (2003). *Case Study Research: Design and Methods.* Thousand Oaks, CA: Sage.

Zarraga, C., & Bonache, J. (2005). The impact of team atmosphere on knowledge outcomes in self-managed teams. *Organization Studies, 26*(5), 661–681.

Zerfaß, A., & Möslein, K. M. (Eds.) (2009). *Kommunikation als Erfolgsfaktor im Innovationsmanagement: Strategien im Zeitalter der Open Innovation.* Wiesbaden: Gabler.

Zhang, H., & Hiltz, S. R. (2003). Factors that influence online relationship development in a knowledge sharing community. *Americas Conference on Information Systems (AMCIS) 2003*, 410–417.

Annexes

Annex A: Open-I platform

The following screenshot displays functionalities associated with innovation development on the Open-I platform.

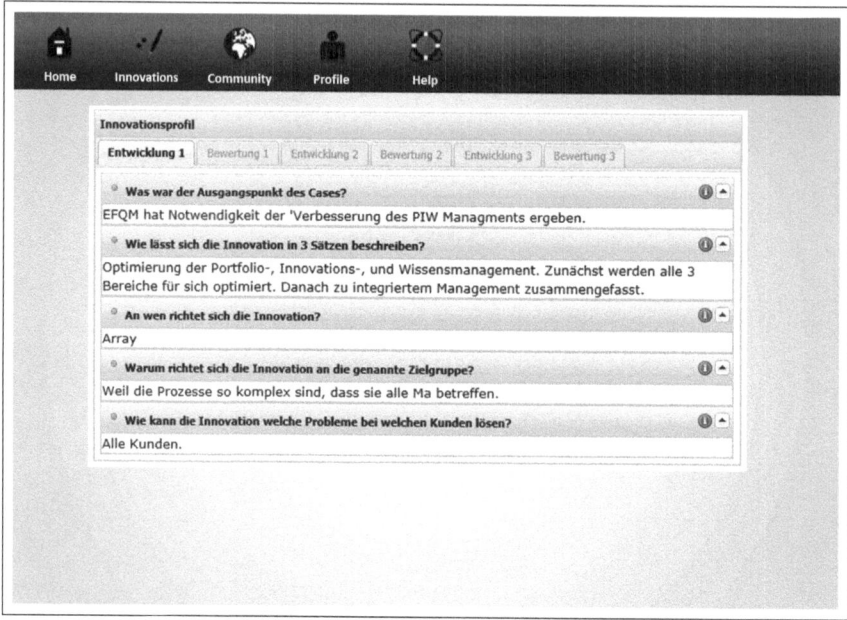

Figure 13: Innovation development on the Open-I platform

In the subsequent screenshot functionalities related to peer based innovation evaluation are displayed.

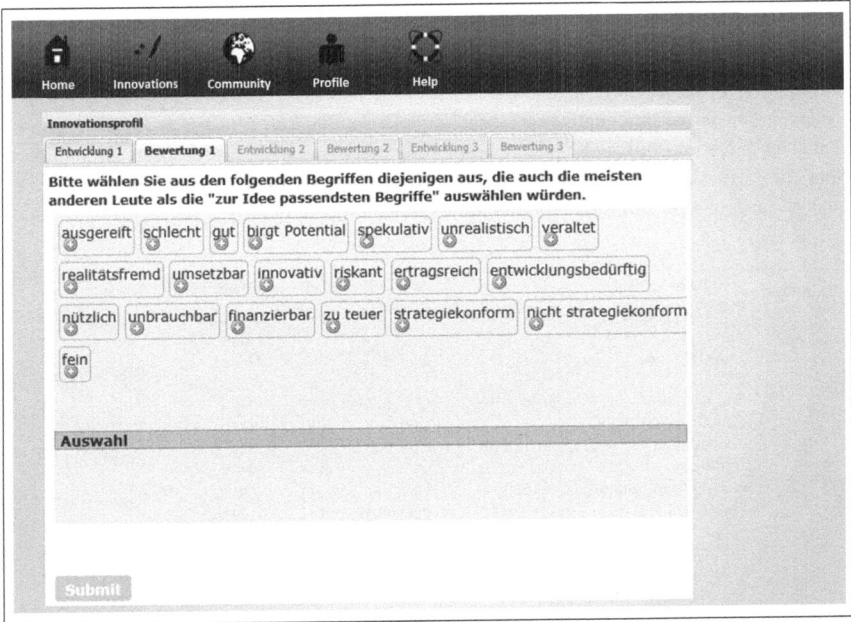

Figure 14: Innovation evaluation on the Open-I platform

Annex B: Contents concerning boundary-less communities

On the *organizational level*, many authors state that ICT is a critical success factor for the process of community building (Case et al., 2001, p. 65; Wamalwa, 2007, p. 79). The main purpose of ICT in boundary-crossing communities is easy and long-term access to information (Case et al., 2001, p. 64; Mican et al., 2009, p. 21) and support for virtual collaboration across time and space (Jäkälä & Pekkola, 2007, p. 12; Case et al., 2001, p. 64; Talukder & Yeow, 2007, p. 82). To achieve these objectives three main aspects are found in literature: 1) aesthetic design, 2) convenient navigation, 3) coordination of social interactions, and 4) constant adaptation.

Main topic:	Related topics:	Applied method:	Focus:	Study:
ICT	Social process, community structure	Descriptive case study	Virtual community	Boczkowski, 1999
	Social process	Conceptual	Online community	Brown, 2002
	Community structure, participation, motivation	Quantitative case study	Online community	Butler et al., 2007
	Community structure, knowledge process	Conceptual	Virtual community	Case et al., 2001
	Knowledge process, community structure	Quantitative and qualitative case study	Virtual community	Chang et al., 2003
	Strength of ties, motivation, participation	Quantitative case study	Virtual community	Ishii & Ogasahara, 2007
	Motivation	Conceptual	Virtual community	Jäkälä & Pekkola, 2007
	Social process	Qualitative case study	Virtual community	Mican et al., 2009
	Community structure	Descriptive case study	Online community	Stanoevska-Slabeva, 2002
	Trust, participation, satisfaction, success	Conceptual	Virtual community	Talukder & Yeow, 2007
	Community structure	Conceptual	Online community	Wamalwa, 2007

Table 63: Publications assigned to organizational context

On the *community level*, several studies show that boundary objects do not only facilitate the creation of shared understanding and knowledge processes, but they also support social processes and the participation of community members (Chang et al., 2007, p. 238; de Cindio et al., 2003, pp. 398, 400, 402; Wachter et al., 2000, p. 481). They are defined as artifacts that perform a brokering role between multiple intersecting social worlds (Star & Griesemer, 1989, p. 393; Star, 1989). Boundary objects in boundary-crossing community literature are classified into two categories: 1) defined rules and 2) common value system.

Additionally, scholars state that *seeding structures* mainly rely on community members take over specific roles to ensure long-term sustainability (Oh & Jeon, 2007, p. 1087; de Cindio et al., 2003, p. 402; Chang et al., 2007, p. 237; Chang et al., 2004, p. 386). They perform certain roles, such as 1) facilitator, 2) administrator, 3) contributor, and 4) developer.

Main topic:	Related topics:	Applied method:	Focus:	Study:
Boundary objects	Knowledge process	Descriptive case study	Virtual community	Chang et al., 2004
	Social process	Qualitative case study	Online community	de Cindio et al., 2003
	Knowledge process, ICT, community structure	Descriptive case study	Virtual community	Wachter et al., 2000
	Strength of ties, participation	Theoretical	Open source community	Oh & Jeon, 2007
Seeding structure	Trust, motivation, status	Descriptive case study	Online community	Cova & Pace, 2006
	Strength of ties, motivation, knowledge process	Qualitative case study	Open source community	Cromie & Ewing, 2008
	Strength of ties	Quantitative case study	Online community	Ganley & Lampe, 2009
	Social process, strength of ties	Quantitative case study	Virtual community	Hummel & Lechner, 2002
	Social process	Conceptual	Virtual community	Kahai et al., 2007
	Social process	Theoretical	Online community	Lin et al., 2009
	Boundary object	Quantitative case study	Open source community	O'Mahony & Ferraro, 2007
	Social process	Social network analysis	Virtual community	Wang & Chen, 2004

Table 64: Publications assigned to community context

On the *individual level*, one of the prevalent factors that greatly influence the vitality of boundary-crossing communities is motivation. Motivation is understood as the activation of individual activity to achieve a certain goal (Rollinson, 2008, p. 196). It is found to include four major motives (Deci & Ryan, 2002; Hars & Ou, 2002, pp. 2, 7; Marett & Joshi, 2009, p. 6; Ke & Zhang, 2009, p. 40; Hertel et al., 2003, p. 1174; Hall & Graham, 2004, p. 240): 1) utilitarian motives, 2) normative motives, 3) collaborative motives, and 4) altruistic motives.

Moreover, trust is widely accepted as a crucial element that defines community coordination for participation and effective interaction (Adler, 2001; Casalo et al., 2008a, p. 334; Ridings et al., 2002, p. 277). Luhmann (1979) characterizes trust as the problem of acting, where the reaction of the exchange partner is not known before. Trust building is positively influenced by ability (judgment), benevolence, and integrity (Porter & Donthu, 2008; Ridings et al., 2006). Four major means of trust are frequently distinguished: 1) dispositional trust, 2) calculus-based trust, 3) information-based trust, and 4) identification-based trust (Abdul-Rahman & Hailes, 2000, p. 102; Leimeister et al., 2005).

Main topic:	Related topics:	Applied method:	Focus:	Study:
Motivation	Performance, participation, status	Quantitative case study	Online community	Beenen et al., 2004
	Participation, outcomes	Quantitative case study	Online community	Franke & Shah, 2003
	Participation	Quantitative case study	Virtual community	Füller, 2006
	Strength of ties, participation	Qualitative interviews	Online community	Hall & Graham, 2004
	Participation	Conceptual	Open source community	Hars & Ou, 2002
	Participation, community structure, status	Quantitative case study	Online community	Hennig-Thurau et al., 2004
	Participation, social process, satisfaction	Quantitative questionnaire	Open source community	Hertel et al., 2003
	Participation, performance, social process	Quantitative questionnaire	Open source community	Ke & Zhang, 2009
	Social process, boundary object	Quantitative questionnaire	Online community	Marett & Joshi, 2009
	Participation, satisfaction	Quantitative questionnaire	Online community	Nonnecke et al., 2006
	Community structure, participation	Qualitative interviews	Open source community	Nonnecke & Preece, 2001
	Trust, participation, seeding structures, outcome	Qualitative case study	Open source community	Shah, 2006
Trust	Social process	Quantitative questionnaire	Online community	Andrews, Preece, & Turoff, 2002
	ICT, social process	Theoretical	Online community	Ba, 2001
	Satisfaction, social process, strength of ties	Quantitative case study	Virtual community	Bauer & Grether, 2005
	Knowledge process, community structure	Theoretical	Virtual community	Birchall & Giambona, 2007
	Social process, strength of ties, trust	Quantitative case study	Online community	Bolton et al., 2004
	Social process	Conceptual	Virtual community	Campbell, Fletcher, & Greenhill, 2007
	Performance, social process	Quantitative questionnaire	Virtual community	Casalo et al., 2008a
	ICT	Conceptual	Virtual community	Castelfranchi & Tan, 2002
	Social process	Quantitative questionnaire	Online community	Feng et al., 2004
	ICT, knowledge process, strength of ties	Descriptive case study	Online community	Kling & Courtright, 2003
	Social process, ICT	Descriptive case study	Virtual community	Leimeister et al., 2005
	Social process, boundary objects, trust, strength of ties	Quantitative and qualitative case study	Virtual community	Mathwick et al., 2008
	Social process, community structure	Quantitative questionnaire	Virtual community	Porter & Donthu, 2008
	Motivation	Quantitative questionnaire	Open source community	Ridings et al., 2002
	Motivation, social process, participation	Quantitative questionnaire	Virtual community	Ridings et al., 2006

Table 65: Publications assigned to individual inputs

On the *process level*, knowledge processes are crucial for the development of boundary-crossing communities. Successful boundary-crossing communities adequately manage knowledge exchange, which make them a superior source for innovation (Hemetsberger & Reinhardt, 2006, p. 210). Within boundary-crossing community literature, knowledge processes predominantly study the creation and distribution of tacit knowledge, i.e., "[…] nonverbalizable, intuitive, unarticulated and context specific in nature" (Polanyi, 1966; Apostolou et al., 2005, p. 14; Hung & Yiyan Li, 2007, p. 487).

Main topic:	Related topics:	Applied method:	Focus:	Study:
Knowledge process	Community structure	Descriptive case study	Virtual community	Apostolou et al., 2005
	Community structure, ICT	Theoretical	Virtual community	Bieber et al., 2002
	Community structure, participation/ knowledge, participation	Theoretical	Open source community	Cheliotis, 2009
	Strength of ties, trust, motivation	Quantitative case study	Virtual community	Chiu et al., 2006
	Social process	Descriptive case study	Open source community	Hemetsberger & Reinhardt, 2006
	Strength of ties	Social network analysis	Virtual community	Hung & Yiyan Li, 2007
	ICT, motivation, satisfaction	Quantitative questionnaire	Online community	Ma & Agarwal, 2007
	Knowledge process, social process, motivation, success	Theoretical	Virtual community	Nambisan, 2002
	Community structure, participation	Quantitative case study, social network analysis	Online community	Wasko, Teigland, & Faraj, 2009
Social process	Motivation	Quantitative case study	Online community	Alonzo & Aiken, 2004
	Strength of ties	Descriptive case study	Open source community	Bergquist & Ljungberg, 2001
	Strength of ties, trust	Conceptual	Virtual community	Blanchard & Markus, 2004
	Trust, participation	Descriptive case study	Virtual community	Donath, 2004
	Motivation, Strength of ties	Quantitative case study	Online community	Jang, Olfman, Ko, Koh, & Kim, 2008
	Community structure	Conceptual	Online community	Kannan, Chang, & Whinston, 2000
	Community structure, participation	Descriptive case study	Online community	Maloney-Krichmar & Preece, 2005
	Strength of ties, community structure	Theoretical	Online community	Panzarasa, Opsahl, & Carley, 2009
	Community structure	Descriptive case study	Online community	Schoberth et al., 2006
	Knowledge process, community structure	Descriptive case study	Online community	Silva, Goel, & Mousavidin, 2009
	Strength of ties	Theoretical	Online community	Szmigin, Canning, & Reppel, 2005

Table 66: Publications assigned to processes

Knowledge processes traverse 1) socialization, 2) articulation, 3) reconciliation, and 4) application (Bieber et al., 2002, p. 13; Hemetsberger & Reinhardt, 2006, p. 189; Nonaka & Takeuchi, 1995). In an ancillary manner, most activities in communities are embedded in social processes. They are characterized by interactivity, or in other words "[...] the extent to which messages in a sequence relate to each other, and especially the extent to which later messages recount the relatedness of earlier messages" (Schoberth et al., 2006, p. 250; Jones & Rafaeli, 2000, p. 1017). Social processes unfold in three major stages: 1) ignition, 2) identification, and 3) confirmation.

On the level of emergent states, satisfaction of community members plays a crucial role for viability in boundary-crossing communities (Lin, 2008, p. 139; Lin & Lee, 2006, p. 484; Sangwan, 2005, p. 5). It determines long-term participation and free word-of-mouth advertising (de Valck et al., 2007, p. 241; Agichtein et al., 2009, p. 3). Satisfaction is a positive or negative evaluation of experiences within boundary-crossing communities (DeLone & McLean, 1992; Gustafsson et al., 2005). Satisfaction is positively influenced by four means: 1) content quality, 2) interaction quality, 3) system quality, and 4) support quality.

Additionally, the participation of community members is a prerequisite for communities to develop, grow and remain viable (Casalo et al., 2008b, p. 21). Participation describes varying levels of intentional social activities of community members (de Valck et al., 2009, p. 192). It is determined by community members' "attitude towards participation [...] perceived pressure from online group members and [...] perceived control over the act of participation" (Füller et al., 2008, p. 615). Three major means of participation are distinguished: 1) pro-active participation, 2) reactive participation, and 3) passive participation.

Lastly, ties between community members are crucial for boundary-crossing communities, as they affect long-term participation (Zhang & Hiltz, 2003, p. 411). Strength of ties "[...] represents the strength of the dyadic interpersonal relationships in the context of social networks" (Money et al., 1998, p. 79). It is defined by 1) weak ties and 2) strong ties, depending on closeness, support, and association as well as intimacy and frequency of exchange (Brown et al., 2007, p. 4; Marsden & Campbell, 1984; Leimeister et al., 2008, p. 368; Granovetter, 1973).

Main topic:	Related topics:	Applied method:	Focus:	Study:
Satisfaction	Knowledge process, community structure	Descriptive case study	Online community	Agichtein et al., 2009
	Social process, community structure	Quantitative questionnaire	Virtual community	de Valck et al., 2007
	Outcome	Quantitative questionnaire	Virtual community	Kuo, 2003
	Performance, ICT	Quantitative questionnaire	Virtual community	Lin, 2008
	Performance	Quantitative case study	Online community	Lin & Lee, 2006
	Satisfaction, participation	Qualitative interviews, quantitative questionnaire	Virtual community	Sangwan, 2005
Participation	Social process, strength of ties, boundary objects	Quantitative case study	Virtual community	Bagozzi & Dholakia, 2002
	Social process, strength of ties	Quantitative interviews	Open source community	Bagozzi & Dholakia, 2006
	Trust, participation, motivation,	Quantitative	Online	Casalo et al.,

	satisfaction	questionnaire	community	2008b
	Social process	Conceptual	Virtual community	Chan et al., 2004
	Social process	Quantitative case study	Virtual community	de Valck et al., 2009
	Social process	Quantitative questionnaire	Online community	Dholakia et al., 2004
	Community structure, social process, performance	Quantitative case study	Online community	Füller et al., 2006
	Knowledge process, motivation, trust	Quantitative case study	Online community	Füller et al., 2008
	Satisfaction, motivation, boundary objects, ICT	Quantitative questionnaire	Online community	Grace-Farfaglia et al., 2006
	Community structure, social process	Social network analysis	Virtual community	Lea et al., 2006
	Social process	Quantitative case study	Online community	Ludford et al., 2004
	Motivation, performance	Conceptual	Virtual community	Romm & Pliskin, 1997
	Motivation, knowledge process, trust	Theoretical	Online community	Tedjamulia et al., 2005
	Motivation, ICT	Reflexive action research	Online community	Vassileva & Sun, 2007
	Motivation, seeding structure	Quantitative questionnaire	Online community	Wang & Fesenmaier, 2004
	Strength of ties, social process, ICT	Quantitative questionnaire	Online community	Wellman et al., 2001
Strength of ties	Social process	Qualitative interviews, social network analysis	Online community	Brown et al., 2007
	Social process, motivation,	Quantitative questionnaire	Virtual community	Koh & Kim, 2003
	Social process	Quantitative case study	Virtual community	Leimeister et al., 2008
	Social process, seeding structure	Quantitative case study	Virtual community	Matei, 2004
	Community structure	Theoretical	Online community	van Alstyne & Brynjolfsson, 2005
	Social process	Conceptual	Online community	Zhang & Hiltz, 2003

Table 67: Publications assigned to emergent states

Measuring *outcomes* is especially important in boundary-less communities to evaluate whether or not a community achieved its goals (Cothrel & Williams, 1999, p. 55). As demonstrated, outcomes are influenced by all before mentioned factors in one way or the other (Leimeister & Krcmar, 2004, p. 9). Outcomes are frequently described in terms of 1) contribution, 2) domains of specialty, and 3) innovation.

Main topic:	Related topics:	Applied method:	Focus:	Study:
Outcomes	Social process, community structure, ICT	Theoretical	Virtual community	Balasubramanian & Mahajan, 2001
	Trust, strength of ties, social process	Quantitative case study	Online community	Chi et al., 2009
	Community structure	Conceptual	Online community	Cothrel & Williams, 1999
	Social process, community structure	Conceptual	Open source community	Dahlander et al., 2008
	Strength of ties, community structure	Descriptive case study	Online community	Farquhar & Rowley, 2006
	Motivation, ICT, trust	Quantitative case study	Virtual community	Ginsburg & Weisband, 2004
	Social process, community structure	Participative action research, social network analysis	Online community	Gloor et al., 2008
	Community structure	Quantitative case study	Open source community	Grewal et al., 2006
	Community structure	Quantitative case study	Virtual community	Gu et al., 2007
	Community structure	Theoretical	Online community	Iriberri & Leroy, 2009
	Knowledge process	Conceptual	Online community	Krieger & Müller, 2003
	ICT, seeding structure	Conceptual	Virtual community	Leimeister & Krcmar, 2004
	Community structure	Quantitative questionnaire	Virtual community	Leimeister et al., 2006
	Participation, boundary object	Descriptive case study	Online community	Otto & Simon, 2008
	Social process	Conceptual	Online community	Preece, 2001
	Social process, knowledge process	Conceptual	Virtual community	Rothaermela & Sugiyamab, 2001

Table 68: Publications assigned to outcomes

Springer Gabler RESEARCH

„Markt- und Unternehmensentwicklung / Markets and Organisations"
Herausgeber: Prof. Dr. Dres. h.c. Arnold Picot,
Prof. Dr. Prof. h.c. Dr. h.c. Ralf Reichwald, Prof. Dr. Egon Franck,
Prof. Dr. Kathrin Möslein
zuletzt erschienen:

Bastian Bansemir
Organizational Innovation Communities
2013. XVIII, 180 S., 14 Abb., 68 Tab., Br. € 59,95
ISBN 978-3-658-01301-1

Kay H. Hofmann
Co-Financing Hollywood Film Productions with Outside Investors
An Economic Analysis of Principal Agent Relationships in the U.S. Motion Picture
Industry
2013. XVIII, 159 S., 10 Abb., 30 Tab., Br. € 59,95
ISBN 978-3-658-00786-7

Ralph Pfaller
IT-Outsourcing-Entscheidungen
Analyse von Einfluss- und Erfolgsfaktoren für auslagernde Unternehmen
2013. XX, 203 S., 31 Abb., 25 Tab., Br. € 49,95
ISBN 978-3-658-00714-0

Jessica Scheler
Driving Innovation in Service Organisations
A Study in the German Airport Industry
2013. XIX, 196 S., 16 Abb., 19 Tab., Br. € 49,95
ISBN 978-3-8349-3406-2

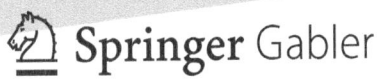
Springer Gabler

Änderungen vorbehalten. Stand: Januar 2013. Erhältlich im Buchhandel oder beim Verlag.
Abraham-Lincoln-Str. 46 . 65189 Wiesbaden . www.springer-gabler.de